Lecture Notes in Computer Science   11885

Pedro Ribeiro · Augusto Sampaio (Eds.)

Unifying Theories of Programming

7th International Symposium, UTP 2019
Dedicated to Tony Hoare on the Occasion of His 85th Birthday
Porto, Portugal, October 8, 2019
Proceedings

 Springer

Editors
Pedro Ribeiro 🄳
University of York
York, UK

Augusto Sampaio
Universidade Federal de Pernambuco
Recife, Brazil

ISSN 0302-9743   ISSN 1611-3349   (electronic)
Lecture Notes in Computer Science
ISBN 978-3-030-31037-0   ISBN 978-3-030-31038-7   (eBook)
https://doi.org/10.1007/978-3-030-31038-7

LNCS Sublibrary: SL1 – Theoretical Computer Science and General Issues

Cover illustration: The cover illustration is the work of Pedro Ribeiro, UK. Used with permission.
Photograph on p. V: The photograph of the honoree was provided by himself. Used with permission.

This Springer imprint is published by the registered company Springer Nature Switzerland AG
The registered company address is: Gewerbestrasse 11, 6330 Cham, Switzerland

Tony Hoare

Preface

This volume contains papers presented at UTP 2019, the 7th International Symposium on Unifying Theories of Programming, held in Porto, Portugal, on October 8, 2019. The event was co-located with the Third World Congress on Formal Methods Europe. Established in 2006, the UTP symposium series aims at bringing together researchers interested in the fundamental problem of the combination of formal notations and theories of programming. The theories define, in various different ways, many common notions, such as abstraction, refinement, composition, termination, feasibility, locality, concurrency, and communication. Despite these differences, such theories may be unified in a way that greatly facilitates their study and comparison. Moreover, such a unification offers a means of combining different languages describing various facets and artifacts of software development in a seamless, logically consistent, way.

This edition of the UTP symposium is in honor of Sir Tony Hoare, on the occasion of his 85th birthday. The papers presented here were invited, and friendly refereed, original contributions sought from the UTP community. One of the papers is from the distinguished invited speaker Tony Hoare himself. Nine other additional papers compose this volume, covering several aspects of Unifying Theories of Programming.

Tony's contributions to Computer Science have been extremely influential. In what follows we revisit some of his major lifetime achievements. A full account can be found in the excellent biographical chapter "Insight, Inspiration and Collaboration" by Jones and Roscoe, published in "Reflections on the Work of C.A.R. Hoare" on the occasion of Tony's 75th birthday in 2009, from which we borrow some biographical information presented in the sequel.

Charles Antony Richard Hoare was born on January 11, 1934, in Colombo, Sri Lanka. After his family returned to England, he attended the Dragon School, Oxford and King's School, Canterbury before going to Oxford University to obtain his highest official academic qualification, an MA in Classical Greats (1952–1956). He studied Latin and Greek, both the languages and their literature. This was followed by Ancient History and Philosophy. It was philosophy that stimulated his interest in logic and the foundations of mathematics. It also stimulated his interest in computers as a tool for studying the philosophy of the mind, and in particular human intelligence.

Tony earned professional qualifications as an interpreter of Russian (1956–58) and as a Statistician (1958–59). He exercised his Russian as an intern at Moscow State University (1959–60), and on many subsequent visits to the Soviet Union (or Russia) and to Armenia. It was in Moscow that Tony is reported to have invented Quicksort. On his return to England in 1960 where he joined the small computer manufacturer Elliott Brothers, he would bet his manager that his algorithm would usually run faster than Shellsort, which Tony had been tasked with implementing for the Elliott 503. He held the posts of Programmer, Chief Engineer, and Researcher (1960–68). He famously led the team (including his wife Jill) that wrote one of the first ALGOL 60 compilers, which stimulated his interest in formal language definition. His experience with

ALGOL led to advocating simplicity in language design, to "enable good ideas to be elegantly expressed" and to "maximise the number of errors which cannot be made, or if made, can be reliably detected at compile time." He then led a much larger failed project to write an operating system for the Eliott 503 Mark II, which stimulated his life-long interest in concurrency.

After Elliott Brothers was taken over in 1968, Tony decided to leave and apply for the Chair of Computing Science at the Queen's University of Belfast, and to his surprise he was appointed (1968–77). He set up the University's first undergraduate degree in Computer Science, at first joint with Mathematics. He selected Program Verification as the topic for his first academic publication, and for his entire subsequent academic research career (thirty-two years). He reckoned that Verification was a topic not likely to be of interest to industry until after he had retired. So he would never have to compete with more generously funded industrial research. And his prediction was quite correct.

In 1977, Tony left Belfast with regret, to return to Oxford University as Professor of Computation (1977–99). There he learned denotational semantics from Joe Stoy and domain theory from Dana Scott. After two years of pure research, he set up a Master's degree at Oxford in Computation, and much later an undergraduate degree (at first joint with Mathematics, then with Engineering).

Tony's first retirement job was in Cambridge with Microsoft Research (1999–2017) where he stimulated an International Grand Challenge Initiative in Verified Software. Since his second retirement, he has been an unpaid visitor at Microsoft Research and an Honorary Member of the Cambridge University Computing Laboratory.

There is no doubt that Tony's contributions to programming have been extremely influential. The "Axiomatic Basis" is one of the most influential papers in Computer Science, which paved the way for reasoning entirely non-operationally about program correctness stepwise. Just as important, the ideas presented in "Structured Programing" have had a major impact on software design.

Concurrency has been a life-long research interest of Tony, having initially been inspired by problems found in operating system design. The beginnings of CSP could be seen emerging in "Parallel Programming: An Axiomatic Approach," where Tony approached shared variable concurrency with his monitor proposal. The first version of "Communicating Sequential Processes," the CACM paper from 1978, proposed an extension of Dijkstra's language of guarded commands with point-to-point communication. The algebra of CSP appeared with the 1985 book of the same name after years of collaborative work with Roscoe and Brookes. CSP would be at the core of the occam programming language, which was prominently used in the context of the transputer, a chip created at inmos. It was used both for programming and as a medium for specification of hardware, serving as a basis for formal verification of the FPU of the transputer, for example. Since then, other programming languages have adopted communication primitives inspired by those of CSP.

Later in 1984, He Jifeng would join Tony in Oxford until 1998. This collaboration led to the extremely ambitious endeavor of creating a relational framework in which a wide range of programming languages, such as imperative, logical, and concurrent, could be given semantics, ultimately culminating in the publication of the book *Unifying Theories of Programming*, which is the *raison d'être* of the UTP symposium series started in 2006. Since then, several contributions have been made to incorporate other aspects, such as: angelic nondeterminism, aspect orientation, event-driven programs, model checking, object orientation, references and pointers, probabilistic programs, real-time programs, reversible computation, synchronicity, timed reactive programs, and transaction processing, to name a few. The UTP has also been embedded in a variety of theorem provers, providing the modern engineer with tools for engineering trustworthy software and systems.

More recently, Tony has revisited algebra as the central unification approach to programming, exploring the interplay between geometrical constructions and algebraic deductions in programming, the interactions and symmetries of space, time, and causality in concurrency. These lie at the foundation of any truly unifying approach to computational modeling and programming. And who, if not Tony, would have the authority to address them?

We are very grateful to Ana Cavalcanti, Jim Woodcock, and Huibiao Zhu for trusting us the honor of organizing the 7th UTP symposium. In doing so we have invited authors of the highest calibre to celebrate Tony's many lifelong contributions to the state of the art in the UTP. We thank all the authors for their contributions, and the Program Committee members for their excellent work in the friendly review process. We are indebted to all the members of the Organizing Committee of the Third World Congress of Formal Methods Europe, led by José Nuno de Oliveira. We are also indebted to EasyChair that greatly simplified the assignment and reviewing of the submissions as well as the production of the material for the proceedings. Finally, we thank Springer for their cooperation in publishing the proceedings.

On his 85th birthday, we salute Tony!

October 2019

Pedro Ribeiro
Augusto Sampaio

Organization

Program Committee

Bernhard K. Aichernig	TU Graz, Austria
Richard Banach	The University of Manchester, UK
Andrew Butterfield	Trinity College Dublin, Ireland
Ana Cavalcanti	University of York, UK
Gabriel Ciobanu	Romanian Academy, Institute of Computer Science, Romania
Hung Dang Van	UET, Vietnam National University, Vietnam
Simon Foster	University of York, UK
Leo Freitas	Newcastle University, UK
Jeremy Gibbons	University of Oxford, UK
Walter Guttmann	University of Canterbury, New Zealand
Ian J. Hayes	The University of Queensland, Australia
Jeremy Jacob	University of York, UK
Zhiming Liu	Southwest University, USA
David Naumann	Stevens Institute of Technology, USA
Marcel Vinicius Medeiros Oliveira	Universidade Federal do Rio Grande do Norte, Brazil
Shengchao Qin	Teesside University, UK
Pedro Ribeiro	University of York, UK
Augusto Sampaio	Universidade Federal de Pernambuco, Brazil
Georg Struth	The University of Sheffield, UK
Burkhart Wolff	Universite Paris-Sud, France
Jim Woodcock	University of York, UK
Naijun Zhan	Institute of Software, Chinese Academy of Sciences, China
Yongxin Zhao	East China Normal University, China
Huibiao Zhu	East China Normal University, China

Contents

Unification Approaches

A Calculus of Space, Time, and Causality: Its Algebra, Geometry, Logic

Tony Hoare[1], Georg Struth[2], and Jim Woodcock[3(✉)]

[1] University of Cambridge, Cambridge, UK
t-tohoar@outlook.com
[2] University of Sheffield, Sheffield, UK
g.struth@sheffield.ac.uk
[3] University of York, York, UK
jim.woodcock@york.ac.uk

Abstract. The calculus formalises human intuition and common sense about space, time, and causality in the natural world. Its intention is to assist in the design and implementation of programs, of programming languages, and of interworking by tool chains that support rational program development.

The theses of this paper are that Concurrent Kleene Algebra (CKA) is the algebra of programming, that the diagrams of the Unified Modeling Language provide its geometry, and that Unifying Theories of Programming (UTP) provides its logic. These theses are illustrated by a formalisation of features of the first concurrent object-oriented language, Simula 67. Each level of the calculus is a conservative extension of its predecessor.

We conclude the paper with an extended section on future research directions for developing and applying UTP, CKA, and our calculus, and on how we propose to implement our algebra, geometry, and logic.

Keywords: Concurrent Kleene Algebra (CKA) ·
Concurrent Separation Logic (CSL) ·
Calculus of Communicating Systems (CCS) ·
Communicating Sequential Processes (CSP) · Action algebra ·
Discrete euclidean geometry · Cartesian coordinates ·
Unified Modeling Language (UML) ·
Unifying Theories of Programming (UTP)

Foreword from Tony

I am deeply grateful to the organisers and attendants at this meeting of UTP 2019 in celebration of my 85th birthday. I could not ask for a better birthday present. I hope you enjoy hearing my presentation as much as I have enjoyed writing it with my co-authors.

Twenty years ago I had the privilege of delivering a presentation at the first World Congress on Formal Methods, FM'99, in Toulouse [74,75]. I gave a talk

P. Ribeiro and A. Sampaio (Eds.): UTP 2019, LNCS 11885, pp. 3–21, 2019.
https://doi.org/10.1007/978-3-030-31038-7_1

entitled "Theories of Programming: Top-down and Bottom-up and Meeting in the Middle" [32]. I claimed that denotational semantics was at the top, operational semantics at the bottom, with algebraic semantics as a unifying link in the middle. My talk today is on the same subject. At the top, discrete geometric diagrams provide the denotations, and its rules of reasoning provide both an operational semantics (e.g., Milner's CCS) and a verification semantics (e.g., O'Hearn's CSL).

Next year is the 60th anniversary of my invention of Quicksort [29], and I propose to retire from active personal research. This will be my farewell appearance at an international conference. I have taken advantage of the opportunity to present a sort of testament, reporting the results of the last ten years of my research. I hope you will find some of them helpful or inspiring for your next ten years of research into the theory and application of UTP.

*   *   *

1   Introduction

The purpose of this paper is to give independent descriptions of the features of programming languages, independently of the language in which they are embedded. It offers many examples, but does not make any recommendation on their selection or rejection. Above all it shuns any attempt to define a complete language. To achieve its purpose, it exploits the power of elementary algebra, geometry, and logic.

The basic insight of the paper is that causality, space, and time have the same meaning inside a computer as in the natural world outside it. These concepts do not require a mathematical semantics, but they will be used to give one.

The operators of sequential composition of operands executed in the same region of space at different intervals of time is an equal partner of concurrent composition of operands executed in the same interval of time in disjoint regions of space. They differ primarily in their domain of definition.

The other insight is that a program is a predicate. Each phrase of a structured program defines exactly, and in minute detail, the set of all its traces of its execution. Each trace records all the events that were performed, both internal to the computer and in interaction with its environment. The execution may be on any computer, at any time and at any place, and with any resolution of its internal non-determinism.

Specifications are predicates that extend the range of operators available in a programming language by including operators of the predicate calculus. Disjunction is most useful in the design phase of development to postpone decisions between design options until more information is available. Conjunction combines requirements in the specification phase of design. In general, its implementation is indescribably inefficient for a non-deterministic program.

Choice in a program is defined by disjunction of predicates, either finite or infinite. An internal choice introduces (demonic) non-determinism into the execution of programs. In making an external choice (for example, by a conditional or a guarded command), the surrounding environment (either the rest of the program or the external world) can prevent selection of one or more of the options.

The trace is written in a subset of the same language as its program, but avoids all forms of choice. It is pictured as an abstract syntax tree (AST): its nodes represent component phrases of a structured program, and each leaf represents a unique execution of a basic commands of the program. Relations between traces are defined by structural induction on their ASTs, particularly the refinement relation and the function mapping each node of a trace to its leaves.

The language of traces contains only composition operators. Their events are just the union of the events of their operands. For example ';' describes sequential execution of its operands at separate instants in time; another operator '|' describes their concurrent execution at separate locations in space.

The operators are distinguished by different constraints on their implementation or on their use. For ';', the implementer of the language must ensure that no event in the first operand has a cause in the second operand. For '|' the user of the language must ensure that the trace contains no cyclic chain of causation between its events.

The operators of the trace calculus are lifted to sets in the usual way. The sets are downward closed wrto the refinement order. Thus if a program describes a trace, it also describes all the refinements of it. In any phase of development, this is what justifies replacement of an abstract program by one of its refinements.

Other operators can be defined by stronger constraints. For example, the CSP operator '→', which separates the guard of a guarded command from its body, requires every event in the body to have a cause in the guard, and to be a cause of every event in the body [31]. The result of a concurrency operation of CSP must not contain a cyclic causal chain, because that would deadlock the implementation. For '|||' in CSP and for separating conjunction in CSL, there must be no causal link between the operands. This obviously prevents both races and deadlocks.

General negation is an incomputable operator. It must be included in a specification language to permit simple descriptions of safety and security by negating a description of what must not happen. But it cannot be included in any general-purpose programming language. It is therefore included in specification languages like UTP [34], CSL [37], and Concurrent Refinement Algebra (a foundation for rely/guarantee reasoning about concurrent programs) [14].

UTP is a special-purpose descriptive logic for specifying traces. It describes primarily the causal links crossing the interfaces between the phrases of a program, and abstracts from the internal events. Each interface is a labelling function from the links of the interface to a value that passed between from its cause to its effect. Examples are the function that represents a region of memory shared between the left and right operand of ';', and the trace of messages passing between concurrent operands, restricted to channels that they share.

2   Extended Summary

2.1   The Calculus

1. An ordering relation is defined on traces by induction on its AST. Its left operand in general requires the same or fewer real processors for execution than the right operand.
2. It specifies a single decision step taken by a timesharing scheduler, which implements concurrency by sequentially interleaving the threads that appear as its operands. Details are given in [39] and [36].
3. The order is a precongruence, i.e., a preorder that makes all operands of the program monotonic (covariant). Its symmetric closure is an equivalence relation that satisfies all the equational axioms of a CKA [37] that are expressible without '+', which means choice in CKA.
4. The algebra is lifted to sets by precongruence closure, standard for converting a precongruence into an order [8,48]. The same lifting is used by Dedekind to lift fractions to real numbers [9]. It is analogous to the equivalence class construction due to Frege and used by Russell in the definition of natural numbers.
5. Choice in a program is defined as set union. The ordering relation on traces lifts to refinement (set inclusion) on programs.
6. Further operators can be defined on sets, both algebraically and by proof rules; for example: iterators (e.g., the Kleene $*$), and residuals for all operators (e.g., weakest prespecifications ($/$) and postspecifications ($\backslash$) [33]), and fixed points [73].
7. A claim that the calculus can be applied to programs is supported by evidence of the large body of program proofs from either the axiomatic proof rules of CSL or the operational rules of CCS [52]. Hoare triples and Milner transitions are given algebraic definitions as a simple refinement in the calculus. The rules of both CSL and CCS are then proved in the calculus. They are three-line proofs in [36].
8. The definitions given above to triples and transitions are the same. By the reflexivity of equality, the theories differ only in notation! Verification logic and operational semantics have been unified in the closest possible way.

The rest of this summary has been included here only for background information. The text summarised has been excluded from the published article for reasons not unconnected with time and space.

2.2   Causality, Space, and Time

1. Causality denotes a familiar relation between events in the real world, and requires no mathematical semantics. Any such semantics (such as that given for Petri nets [62,63]) can be considered as a scientific theory applicable to the natural world. Causality is represented graphically by drawing an arrow from the causing event to the caused event (its effect) at its head.

2. The essential property of causation is that no event can occur before its cause. But they can occur at the same time.
3. The collection of arrows between the events of a trace forms a directed graph, with events represented by points.
4. Arrows are classified as either vertical or horizontal. A vertical arrow is drawn between the successive events that occurred at the same location in space; events are performed by an object allocated at a given location of memory, for example a variable or a communication port of a channel.
5. A horizontal arrow is drawn between simultaneous events, each performed by a different object (e.g., a thread and a variable). The full set of simultaneous actions is known as a transition [62], or as a transaction [25].
6. The graph for a trace is segmented into subgraphs, one for each node and for each leaf of its AST. A leaf is the only occupant of its segment.
7. The graph of a sequential node is split by a horizontal cut between its operands. The only coordinates that it cuts are vertical. A concurrent node is split by a vertical cut, which cuts only horizontal coordinates.
8. The graph for any node of the AST is contained within a rectangular box, whose edges are cuts. The input arrows of the box are defined as those with only their heads in the box; and those with only tails are called its outputs. Its internal arrows have both ends in the box.
9. Conventionally, all vertical input arrows enter the box at its top edge, and vertical output arrows leave at the bottom edge. Horizontal input arrows enter at the left edge and horizontal output arrows leave it at the right.

2.3 Geometry

1. For purposes of program debugging, a box can be displayed as a diagram of discrete plane geometry. Its two axes represent time and space, and its points represent events.
2. A vertical coordinate is a chain of arrows containing the complete history of events performed by a single object. Examples of objects include variables, threads, communication ports, messages.
3. A horizontal coordinate is a set of arrows connecting all its points. Each point on it is shared by a distinct vertical coordinate. Examples include multiple assignments, communications, synchronising fences, object allocation, and disposal.
4. Any pair of vertical or of horizontal coordinates is mutually parallel in the sense of Euclid: they have no point in common.
5. As in Cartesian geometry, every point in a diagram is the unique element of the intersection of a horizontal and a vertical coordinate.
6. Further examples are presented in [35] and alternatively in [54]. They are called Sequence Diagrams in UML [59], or Message Sequence Charts in SDL [43].
7. In a debugging tool, each error revealed by the trace should be highlighted. The tool should also provide a means for navigating backwards from an error, travelling along vertical and horizontal coordinates to its direct and indirect

causes (time-travel debugging [51]); also forwards to its results whenever possible after a non-fatal error.

2.4   Logic

1. According to its standard semantics, a proposition of predicate calculus describes all observations (aka, valuations) that satisfy it. An observation is a total function from all syntactically possible variable identifiers to their observed values.
2. For the predicates of CSL and of UTP, the observations are only partial functions, whose domain (aka footprint) is the set of all free variables of the proposition. Negation of an assertion preserves its footprint.
3. Separating conjunction in CSL describes concurrent composition of programs. It is defined only if the footprints of its two operands are disjoint, and neither is undefined. The footprint of a disjunction is the union of that of footprints of its operands.
4. The footprint of a predicate in UTP is defined in terms of the box diagrams of the trace described. For its top edge, it is the collection of unique names of the vertical coordinates that cross a horizontal boundary of the box diagram. For the bottom edge of the diagram, the names are annotated by a dash.
5. The value of an object observed at these edges is that which was assigned (or left unchanged) by the event at the tail of the cut arrow.
6. In the Circus variant of UTP [61, 76], a built-in variable tr stands for a record of the history of all input-output events that are recorded from the beginning of the entire trace. Each input of a message lies on the same horizontal interface as the output of the message.
7. Alternative conventions are often more intuitive. For example, the trace can be represented by a finite state diagram in which the nodes are annotated by an invariant that describes the values of the variables throughout the interval between its initial and its final horizontal coordinate.

3   The Calculus

Our terms are traces of execution of a program written in a language that includes events (to be defined later), a constant **1**, and two binary operators of sequential composition ('; ') and concurrent composition ('|'). The context-free syntax of the terms of the calculus is:

$$\langle term \rangle \quad ::= \mathbf{1} \mid \langle event \rangle \mid (\langle term \rangle \langle operator \rangle \langle term \rangle)$$
$$\langle operator \rangle ::= \text{'; '} \mid \text{'|'}$$

By structural recursion we define the events of a term to be the set of events recorded in the whole trace:

$$events(\mathbf{1}) = \{\}$$
$$events(k) = \{k\} \qquad\qquad\qquad\qquad\qquad \text{[if } k \text{ is an event]}$$

$$events(p\,;q) = events(p\,|\,q) = events(p) + events(q)$$

where $\{k\}$ is the singleton set containing only k, and $+$ is disjoint union of sets, ensuring that each event only occurs once in the abstract syntax tree. If the operands of $+$ are not disjoint, then the result is undefined. This fact is expressed by the *ok* predicate of UTP [34], which satisfies the axioms

$$ok(s + t) \equiv (s \cap t = \{\}) \wedge ok(s) \wedge ok(t)$$
$$ok(s) \qquad\qquad\qquad\qquad\qquad \text{[if } s \text{ is a singleton set or empty]}$$

Let p be a term $(r\,;s)$ and let e be a member of $events(r)$ and let f be a member of $events(s)$. Then the pair (e, f) is said to be sequentially separated within p. The set of pairs within p that are so separated is defined by

$$ssep(p) = \{events(r) \times events(s) \mid r\,;s \text{ is a subterm of } p \text{ (or } p \text{ itself)}\}$$

This can also be defined axiomatically without set notation by structural recursion:

$$ssep(1) = \{\}$$
$$ssep(p\,;q) = ssep(p) + ssep(q) + events(p) \times events(q)$$
$$ssep(p\,|\,q) = ssep(p) + ssep(q)$$

A definition of concurrent separation $csep(p\,|\,q)$ is similar.

The structure of two terms is compared by a relation $\leq$ between them.[1] It means that q has a denser sequential control structure than p. For example, $p\,;q \leq p\,|\,q$. (Similar relations can be defined with respect to *csep*.)

$$p \leq q \;\; \widehat{=} \;\; ssep(p) \subseteq ssep(q) \wedge events(p) = events(q)$$
$$p \equiv q \;\; \widehat{=} \;\; (p \leq q) \wedge (q \leq p)$$

This definition of the fundamental ordering relation is formulated in terms of syntax. This may violate one tradition of programming language semantics: that syntax and semantics should be totally separated. The syntactic definition is strongly welcomed in other traditions. It gives the strongest possible model of the calculus. Algebraists call it a word algebra, category theorists call it an initial or free algebra, and computer scientists call it fully abstract. A proof that CKA satisfies all these definitions is given in [47]. A concept with three or more equivalent definitions is usually important in mathematics, for example the axiom of choice in logic.

---

[1] In UTP [17], refinement between pointwise relations is written as $P \sqsubseteq Q$ (or equivalently $Q \leq P$), and defined by $[Q \Rightarrow P]$. It asserts that every behaviour of Q is also a behaviour of P.

3.1   The Algebra of Traces

The axioms of the calculus are just those basic axioms of CKA [37] that can be expressed in the syntax; those involving choice (written as +), repetition (*) and residuation (/ and \) are omitted. They will be re-introduced shortly.

Theorem 1

1. $\leq$ *is a preorder*   [reflexive and transitive]
2. *If $q \leq p$ then $p \mid r \leq p \mid r$ and $r \mid q \leq r \mid p$*   [monotonicity]
         and $p\,;r \leq p\,;r$ and $r\,;q \leq r\,;p$
3. $(p\,;q)\,;r \equiv p\,;(q\,;r)$ *and* $(p \mid q) \mid r \equiv p \mid (q \mid r)$ [associativity]
4. $p\,;1 \equiv p \equiv 1\,;p$ *and* $p \mid 1 \equiv p \equiv 1 \mid p$ [unit]
5. $(p \mid q)\,;(p' \mid q') \leq (p\,;p') \mid (q\,;q')$ [interchange]

The first two laws echo the familiar laws for equality, formulated by Euclid and Leibniz. They permit a refinement to be used as a single-directional substitution rule in algebraic reasoning. A standard structural induction from the second law says that refinement is preserved when the rule is applied to any sub-term of a given term. The third law allows redundant brackets to be omitted. And the fourth describes the steps that reduce a term to sequential normal form, in which all '$\mid$' are eliminated.

We obtain four small interchange laws from Theorem 1.5 by substituting units for each of the four variables.

$$p\,;(r \mid s) \leq (p\,;r) \mid s \qquad q\,;(r \mid s) \leq r \mid (q\,;s)$$

$$(p \mid q)\,;s \leq p \mid (q\,;s) \qquad (p \mid q)\,;r \leq (p\,;r) \mid q$$

Two tiny interchange laws are derived by a second such substitution in the first line above:

$$p\,;r \leq p \mid r \qquad q\,;r \leq r \mid q$$

The interchange axiom models the decisions of a timesharing scheduler operating at run time or at compile time. Its purpose is to reduce the number of actual processors needed for execution of a program below what it has explicitly called for. In combination with the equational axioms, it may be used as a single step in the reduction of any term of the calculus to a normal form that has no '$\mid$'. The equational axioms are used first on each step to select which '$;$'s and which '$\mid$'s to match to the left hand side of interchange. Different choices will result in different eventual interleavings. Each non-trivial application of interchange increases the membership of *sseq*, so the shuffling process must terminate. The corollaries of the axiom are what finally eliminates the '$\mid$'s.

3.2 Applications

The simplicity, relevance, and power of the calculus is demonstrated by its application to two well known and widely used theories of programming, separation
logic [60, 66] (which includes Hoare Logic) and Milner's CCS [52]. The Hoare
triple $\{p\}\, q\, \{r\}$ [30] is interpreted as saying that performance of q preceded by
p is one of the ways of implementing r: i.e., $p\,;q \leq r$. (This is a generalisation
of the original Hoare definition, which required that p and r be restricted to the
events that evaluate assertions [72], in a similar manner to weakest prespecifications [33].) From this definition, the proof rules for sequencing and concurrency
in CSL (Concurrent Separation Logic) [38]. Simpler proofs (three-liners mostly)
are given in [38].

The Milner transition is written $r \xrightarrow{p} q$. In the small-step version of the
transition, the program p is restricted to a singleton event. This triple is interpreted as the statement that one of the ways of implementing r is to perform p
first, saving its continuation q for later execution. Algebraically expressed, this
is $p\,;q \leq r$, which is the same definition as the Hoare triple. By definition, the
two calculi are the same! This claim can be checked by definitional substitution,
which translates the defining axioms of each theory into those of the other. The
unification is similar to that made by Dirac, when he showed the mathematical
identity of the Schrödinger and the Heisenberg formulations of quantum theory
with his own.

3.3 The Algebra of Programs

The behaviour of a program is defined as the set of all traces that can be produced
by its execution. The operators are defined by complex product or convolution.
Let capital letters stand for sets of traces. Define the operators on the traces by

$$
\begin{aligned}
1 \quad &\hat{=} \{1\} \\
P\,;Q \quad &\hat{=} \{(p\,;q) \mid p \in P \land q \in Q\} \\
P\,|\,Q \quad &\hat{=} \{(p\,|\,q) \mid p \in P \land q \in Q\} \\
P \sqcap Q &\hat{=} P \cup Q
\end{aligned}
$$

Nondeterministic choice is defined as set union, and its algebraic properties are
familiar from Boolean algebra: it is associative, commutative, and idempotent,
and it has the empty set as its unit. Refinement is defined by set inclusion.

Linearity of the axioms in Theorem 1 ensures that the equational properties
of traces remain unchanged when they are lifted to sets of terms in the usual
way, by the results of [20]. We would also like to remove all undefined terms from
the sets. This is done by applying a familiar algebraic construction for turning a
preorder into a partial order, namely by the downward closure of the sets, with
respect to the preorder:

$$
\begin{aligned}
P\,;Q \quad &\hat{=} \{r \mid p \in P \land q \in Q \land r \land ok(r) \leq p\,;q\} \\
P\,|\,Q \quad &\hat{=} \{r \mid p \in P \land q \in Q \land r \land ok(r) \leq p\,|\,q\} \\
P \leq Q &\hat{=} P \subseteq Q
\end{aligned}
$$

Further operators can be defined on sets, both algebraically and by proof rules; for example: iterators (e.g., the Kleene $*$), and residuals for all operators (e.g., weakest prespecifications (/)and postspecifications (\) [33]), and fixed points [73]. Iteration is defined as the least fixed point of $x = SKIP \vee x \vee x \, ; x$. An introduction to these topics is well presented by axioms and proof rules in [64], where a complete algebraic characterisation of iteration includes the elegant equation $p = (p/p)^*$, or equivalently $p = (p \setminus p)^*$, where p is an invariant of the loop. Pratt proved the axiom interdeducible with the proof rules that define either least or greatest fixed-point (depending on the order).

4 Symmetries

In the natural sciences, an experiment is designed to produce a result that all observers of a repeated experiment will agree on, no matter when and no matter where it is viewed from. The raw observations will obviously be different, but agreement can be reached if the direct description of each raw observation is automatically translatable into a description made by any observer from a different viewpoint at a later time. The translation algorithm is called a symmetry.

It is therefore not surprising that the laws themselves are translatable by the same symmetry, and each translation to gives back either the same law or another one. That gives confidence of the universal applicability of the laws throughout space and time. It should certainly be checked by mathematical proof.

The axioms in Theorem 1 satisfy such symmetries. In the algebra, we model time reversal symmetry by a function v that swaps the arguments of '$;$', space inversion symmetry by a function h that swaps those of '$|$', and space-time symmetry by a function d that interchanges '$;$' and '$|$':

$$v(p\,;q) = v(q)\,;v(p), \qquad v(p\,|\,q) = v(p)\,|\,v(q) \qquad v(1) = 1$$
$$h(p\,;q) = h(p)\,;h(q), \qquad h(p\,|\,q) = h(q)\,|\,h(p) \qquad h(1) = 1$$
$$d(p\,;q) = d(p)\,|\,d(q), \qquad d(p\,|\,q) = d(p)\,;d(q) \qquad d(1) = 1$$

As before, '$|$' need not commute. All axioms in Theorem 1 are closed under these symmetries. We explain only the interchange law, which we write as

$$\frac{p}{q}\,;\frac{r}{s} \leq \frac{p\,;r}{q\,;s}$$

Applying the symmetries yields

$$\frac{v(r)}{v(s)}\,;\frac{v(p)}{v(q)} \leq \frac{v(r)\,;v(p)}{v(s)\,;v(q)} \qquad \frac{h(q)}{h(p)}\,;\frac{h(s)}{h(r)} \leq \frac{h(q)\,;h(s)}{h(p)\,;h(r)} \qquad \frac{d(p)\,;d(q)}{d(r)\,;d(s)} \geq \frac{d(p)}{d(r)}\,;\frac{d(q)}{d(s)}$$

The first law holds because *ssep* guarantees that the left-hand and right-hand sides of the two laws have the same points and the same arrows, which are just reversed in direction. The second one holds by a similar argument with respect to *csep*. The third law is valid because d interchanges *ssep* with *csep*, after which the value of *ssep* decreases from right to left. This explains the reversal of $\leq$. Application of d to the monotonicity axioms in Theorem 1 requires this reversal as well.

If p, q, r and s are leaves of the AST, then we can depict the nodes in both sides of the first interchange law as the following black square.

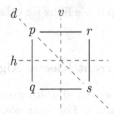

In the symmetric interchange laws, v then reflects the nodes in the vertical axis bisecting the square, h in the horizontal axis, and d in the diagonal axis through p and s. In fact, h can be generated from d and v as $h = d \circ v \circ d$ and the full symmetry group of the square—reflection in the other diagonal and rotations by 0°, 90°, 180° and 270°—can be generated from these two elements as well. Interchange is therefore invariant under all eight symmetries.

In mathematics, symmetries are admired for their beauty. They arise as properties invariant under some transformation, usually the action of some group. Yet beyond their beauty, they have practical uses too. The symmetries of interchange are preserved by refinement, so any conjecture that does not preserve symmetry can be instantly rejected. Furthermore, every theorem proved automatically generates seven corollaries. Exploitation of symmetries by a proof tool can give further optimisations [13, 21, 41, 68].

5   Future Directions

In the shortest term, the authors plan to publish a journal version of this paper with missing sections restored. It is proposed to apply the syntactic methods of this paper to define features like probabilistic choice and delay commands that are found in Simula 67 [7].

An urgent development of the theory presented here is to model the layers of abstraction that are implemented by a hierarchy of class declarations in an object-oriented language. A layer includes all the subclasses of a class, and shares no resources with any other layer. It is sometimes called a component or a module. Abstraction scales: the very largest systems in worldwide use today could never have evolved without it.

5.1   Unifying Theories of Programming

The major problem facing verification today is that many large systems are written in a combination of languages:

- General purpose, application-oriented (e.g., scripting, discrete-event simulation, network design, security).
- Continuous control in both signal-oriented and equational styles.
- Hardware-oriented (e.g., GPU, FPGA, quantum).

How can we provide a common toolset for them all? Perhaps the algebraic methods introduced in this paper could be used to develop a semantics for existing and future languages, with compatible links for other languages used in the same product.

5.2   Applications of Concurrent Kleene Algebra

Kleene algebra is well known for vast simplifications and generalisations of the proofs of some important theorems. For example, in [45], Kozen gives a completely algebraic proof in KA with tests (KAT) that a program with nested loops can be reduced to a program with just a single loop and some auxiliary variables (a classic folk theorem), and von Wright gives a very elegant, single-page proof of a theorem for atomicity refinement in action systems [78] that previously had taken Back many pages to accomplish [2]. Equally convincing results for concurrent programs with CKA are so far missing. Examples would include concurrency control or concurrency refinement laws.

KAT [45], Kleene algebras with domain [10,11], and demonic refinement algebras [78] have been established as abstract semantics and verification methods for sequential programs and linked with concrete program semantics such as relations or predicate transformers. Hayes and co-workers have recently developed concurrent refinement algebras, which are inspired by CKA, and support rely-guarantee style reasoning with shared-variable concurrent programs [27,28] and CCS/CSP-style reasoning [31,52]. Similar applications in the semantics of concurrent programming languages remain to be explored. Many of the approaches mentioned have led to verification components with interactive theorem provers, notably with Isabelle/HOL [1,18,24]. For CKA, such components are under development.

5.3   Implementing the Calculus

In the immediate future, we are planning that a research team at York will engage in developing a library of theories in our theorem prover, Isabelle/UTP [17,19,77], which is an implementation of UTP in Isabelle/HOL [58].

We hope to recruit collaboration with other centres of excellence to develop compatible extensions of the mechanisation in other proof tools, for example Coq [3], Lean [55], Maude [6], Agda [4], and FDR [23].

We will support the geometric presentation of the calculus using Eclipse [71], defining the abstract graphical syntax with the Eclipse Modeling Framework (EMF), its concrete syntax with the Graphical Modeling Framework (GMF), and transforming the models created with the language into Isabelle/UTP theories using the Epsilon model transformation tool [44]. This will follow the approach set out in [53,79], where the graphical RoboChart language [53] is managed within the Eclipse-based RoboTool environment [67] and transformed to CSP [31], PRISM [46], and Isabelle/UTP [19].

5.4 Object Orientation and UTP

Object-oriented programming is the only known programming paradigm that makes writing massive software applications reasonably manageable, maintainable, and scalable. The research presented in this paper uses classes and their objects as the principal technique for abstraction. A full treatment of object orientation requires additional abstraction techniques that provide encapsulation and information hiding, supporting structuring and re-use of classes through inheritance (perhaps including multiple inheritance), behavioural subtyping, and polymorphism; it would permit the use of dynamic dispatch as a way of selecting different implementations.

There is already much significant work on OO in UTP, but an elegant and integrated treatment in UTP remains a significant ambition. Existing achievement include the following. Santos et al. [69] present a general theory of object orientation in UTP. Naumann et al. [57] give a semantics to class hierarchies and how to refactor them for representation independence. Cavalcanti et al. [5] report on unifying OO classes and CSP-like processes in *OhCircus*, an object-oriented extension of the UTP-based *Circus* multi-paradigm language [61,76], with a formalisation of method calls and their refinement. Ramos et al. [65] give a semantics to active classes in UML-RT, the real-time profile for UML, via a mapping into *Circus*. Duran et al. [12] present a strategy for compiling classes, inheritance, and dynamic binding, following the compilation strategy for Dijkstra's guarded command language using refinement algebra in UTP [34, Chap. 6]. Silva et al. [70] present the laws of programming for object orientation with reference semantics and Gheyi et al. [22] give a complete set of object modelling laws for Alloy [42]. Finally, Zeyda and his colleagues [80] present a modular theory of object orientation in higher-order UTP [34, Chap. 9], all mechanised in Isabelle/UTP [17,19,77].

A huge challenge is to harmonise and extend these existing UTP theories to provide a simple and widely accepted treatment of all the main features of object orientation.

To test and evaluate the theory of classes, other concurrent programming design patterns should be specified experimentally as class declarations. At each layer, the programmer needs a way of specifying new behavioural type systems checkable at compile time and proof systems detectable at run time. Their purpose is to avoid violations of the protocols whose universal observance by user programs is required by the design pattern. Type inference algorithms should be specifiable within the algebra, perhaps by restricting refinement rules to Horn clauses [40]. They can then be directly executed by exhaustive tree search. The same restriction is also made in functional languages, but other syntactic restrictions ensure determinacy, so that tree search is not necessary.

5.5 Extensions of the Calculus

Probabilistic Kleene Algebra (PKA) [49] and CKA [37] have been combined in Concurrent Probabilistic Kleene Algebra (CPKA) to provide a unified account of nondeterminism, probability, and concurrency, with models in probabilistic automata, modulo probabilistic refinement simulation [50]. This is a natural target for the extension of the algebra, geometry, and logic of our calculus.

A particular application area of great current interest is Cyber-Physical Systems (CPS). They use embedded computers and networks to compute, communicate, and control physical processes. Research in verification in this area has to provide the techniques and tools for checking the correctness of software and hardware platforms with respect to agreed requirements.

The notion of correctness has to be judged against runtime feedback on the validity of assumptions about the environment, and digital twin technology is being proposed to handle this problem [26]).[2] Fitzgerald et al. [15] describe the beginnings of a generalised theory of CPS design, with an introduction to the formal foundations, methods, and integrated tool chains for CPS. Crucially, models of CPS are inherently heterogeneous and require unification of different languages, design methods, and verification techniques and their tools.

Modal Kleene Algebra (MKA) [10, 11] has recently been used with ordinary differential equations (ODEs) for the verification of hybrid systems, where discrete imperative program behaviour complements continuous physical dynamics [56]. Foster et al. [16] describe a generalisation of the UTP theory of reactive processes [34, Chap. 7] using abstract trace algebra. This extends the reactive process theory to continuous time traces, where events are replaced by piece-wise continuous functions of physical behaviour, and this gives a model of hybrid systems. A connection between the UTP and MKA is a long-term and very ambitious objective.

6 Conclusion

The long-term practical goal of a theory of programming is to provide a conceptual framework for the design of a coherent set of practical tools for program

---

[2] In this extension of model-based engineering, a digital twin is a virtual model of the system, constructed from formal development artefacts and used throughout the lifetime of the product. This pairing of the virtual and physical worlds allows analysis of data and monitoring of systems to detect problems before they occur, prevent downtime, develop new application opportunities, and plan immediate and long-term behaviour using simulations. Since the virtual model captures the assumptions made about the environment during system development, these assumptions can be tuned to more accurately reflect reality.

development. They should cover the features of modern general-purpose programming languages, and also special-purpose languages and design patterns that exploit synergy in the characteristics of particular applications, algorithms and hardware. The tools should cover the entire life cycle of large-scale program evolution, which starts from requirements and specifications, and continues through system architecture, program design, coding, static checking, compilation, optimisation, selective verification, testing, and correction, right up to delivery of the product. The cycle then repeats in subsequent evolution of the delivered product. The coherence of the theory enables the various languages to be used together in the same software architecture. The conceptual framework should ideally be accompanied by tools which give assistance in the life cycle of new special-purpose programming languages likely to emerge in the changing world.

It is comforting that the conceptual framework of causality, space, and time is the same as that of our common-sense world, and of the more advanced theories of modern science.

Acknowledgements. Parts of this work were funded under EPSRC grants EP/R032351/1 on *Verifiably Correct Transactional Memory* and EP/M025756/1 on *A Calculus for Software Engineering of Mobile and Autonomous Robots*, and by a Royal Society grant on *Requirements Modelling for Cyber-Physical Systems*.

References

1. Armstrong, A., Gomes, V.B.F., Struth, G.: Building program construction and verification tools from algebraic principles. Formal Asp. Comput. **28**(2), 265–293 (2016)
2. Back, R.J.R.: A method for refining atomicity in parallel algorithms. In: Odijk, E., Rem, M., Syre, J.-C. (eds.) PARLE 1989. LNCS, vol. 366, pp. 199–216. Springer, Heidelberg (1989). https://doi.org/10.1007/3-540-51285-3_42
3. Bertot, Y., Castran, P.: Interactive Theorem Proving and Program Development: Coq'Art The Calculus of Inductive Constructions. Springer, Heidelberg (2010)
4. Bove, A., Dybjer, P., Norell, U.: A brief overview of Agda—a functional language with dependent types. In: Berghofer, S., Nipkow, T., Urban, C., Wenzel, M. (eds.) TPHOLs 2009. LNCS, vol. 5674, pp. 73–78. Springer, Heidelberg (2009). https://doi.org/10.1007/978-3-642-03359-9_6
5. Cavalcanti, A., Sampaio, A., Woodcock, J.: Unifying classes and processes. Softw. Syst. Model. **4**(3), 277–296 (2005)
6. Clavel, M., Durán, F., Eker, S., Lincoln, P., Martí-Oliet, N., Meseguer, J., Talcott, C.: All About Maude—A High-Performance Logical Framework, How to Specify, Program and Verify Systems in Rewriting Logic. LNCS, vol. 4350. Springer, Heidelberg (2007). https://doi.org/10.1007/978-3-540-71999-1
7. Dahl, O., Myhrhaug, B., Nygaard, K.: Simula 67 common base language. Technical report. NCC, May 1968
8. Davey, B.A., Priestley, H.A.: Introduction to Lattices and Order. Cambridge University Press, Cambridge (1990)
9. Dedekind, R.: Stetigkeit und irrationale Zahlen. Verlag von Friedrich Vieweg und Sohn, Braunschweig (1872)

10. Desharnais, J., Möller, B., Struth, G.: Kleene algebra with domain. ACM Trans. Comput. Log. **7**(4), 798–833 (2006)
11. Desharnais, J., Struth, G.: Internal axioms for domain semirings. Sci. Comput. Program. **76**(3), 181–203 (2011)
12. Duran, A., Cavalcanti, A., Sampaio, A.: A strategy for compiling classes, inheritance, and dynamic binding. In: Araki, K., Gnesi, S., Mandrioli, D. (eds.) FME 2003. LNCS, vol. 2805, pp. 301–320. Springer, Heidelberg (2003). https://doi.org/10.1007/978-3-540-45236-2_18
13. Emerson, E.A., Sistla, A.P.: Symmetry and model checking. Formal Methods Syst. Des. **9**(1/2), 105–131 (1996)
14. Fell, J., Hayes, I.J., Velykis, A.: Concurrent refinement algebra and rely quotients. Archive of Formal Proofs 2016 (2016)
15. Fitzgerald, J.S., Gamble, C., Larsen, P.G., Pierce, K., Woodcock, J.: Cyberphysical systems design: formal foundations, methods, and integrated tool chains. In: Gnesi, S., Plat, N. (eds.) 3rd IEEE/ACM FME Workshop on Formal Methods in Software Engineering, FormaliSE 2015, Florence, 18 May 2015, pp. 40–46. IEEE Computer Society (2015)
16. Foster, S., Cavalcanti, A., Woodcock, J., Zeyda, F.: Unifying theories of time with generalised reactive processes. Inf. Process. Lett. **135**, 47–52 (2018)
17. Foster, S., Woodcock, J.: Unifying theories of programming in Isabelle. In: Liu, Z., Woodcock, J., Zhu, H. (eds.) Unifying Theories of Programming and Formal Engineering Methods. LNCS, vol. 8050, pp. 109–155. Springer, Heidelberg (2013). https://doi.org/10.1007/978-3-642-39721-9_3
18. Foster, S., Ye, K., Cavalcanti, A., Woodcock, J.: Calculational verification of reactive programs with reactive relations and Kleene algebra. In: Desharnais, J., Guttmann, W., Joosten, S. (eds.) RAMiCS 2018. LNCS, vol. 11194, pp. 205–224. Springer, Cham (2018). https://doi.org/10.1007/978-3-030-02149-8_13
19. Foster, S., Zeyda, F., Woodcock, J.: Isabelle/UTP: a mechanised theory engineering framework. In: Naumann, D. (ed.) UTP 2014. LNCS, vol. 8963, pp. 21–41. Springer, Cham (2015). https://doi.org/10.1007/978-3-319-14806-9_2
20. Gautam, N.D.: The validity of equations of complex algebras. Archiv für mathematische Logik und Grundlagenforschung **3**(3), 117–124 (1957)
21. Gent, I.P., Petrie, K.E., Puget, J.: Symmetry in constraint programming. In: Rossi, F., van Beek, P., Walsh, T. (eds.) Handbook of Constraint Programming, Foundations of Artificial Intelligence, vol. 2, pp. 329–376. Elsevier (2006)
22. Gheyi, R., Massoni, T., Borba, P., Sampaio, A.: A complete set of object modeling laws for Alloy. In: Oliveira, M.V.M., Woodcock, J. (eds.) SBMF 2009. LNCS, vol. 5902, pp. 204–219. Springer, Heidelberg (2009). https://doi.org/10.1007/978-3-642-10452-7_14
23. Gibson-Robinson, T., Armstrong, P.J., Boulgakov, A., Roscoe, A.W.: FDR3: a parallel refinement checker for CSP. STTT **18**(2), 149–167 (2016)
24. Gomes, V.B.F., Struth, G.: Modal Kleene algebra applied to program correctness. In: Fitzgerald, J., Heitmeyer, C., Gnesi, S., Philippou, A. (eds.) FM 2016. LNCS, vol. 9995, pp. 310–325. Springer, Cham (2016). https://doi.org/10.1007/978-3-319-48989-6_19
25. Gray, J., Reuter, A.: Transaction Processing: Concepts and Techniques. Morgan Kaufmann, Burlington (1993)
26. Grieves, M., Vickers, J.: Digital twin: mitigating unpredictable, undesirable emergent behavior in complex systems (excerpt). Technical report. University of Michigan, August 2016

27. Hayes, I.J.: Generalised rely-guarantee concurrency: an algebraic foundation. Formal Asp. Comput. **28**(6), 1057–1078 (2016)
28. Hayes, I.J., Meinicke, L.A., Winter, K., Colvin, R.J.: A synchronous program algebra: a basis for reasoning about shared-memory and event-based concurrency. Formal Asp. Comput. **31**(2), 133–163 (2019)
29. Hoare, C.A.R.: Algorithm 64: quicksort. Commun. ACM **4**(7), 321 (1961)
30. Hoare, C.A.R.: An axiomatic basis for computer programming. Commun. ACM **12**(10), 576–580 (1969)
31. Hoare, C.A.R.: Communicating Sequential Processes. Prentice Hall, Upper Saddle River (1985)
32. Hoare, C.A.R.: Theories of programming: top-down and bottom-up and meeting in the middle. In: Wing, et al. [74], pp. 1–27
33. Hoare, C.A.R., He, J.: The weakest prespecification. Inf. Process. Lett. **24**(2), 127–132 (1987)
34. Hoare, C.A.R., He, J.: Unifying Theories of Programming. Prentice Hall, Upper Saddle River (1998)
35. Hoare, T.: Geometric theory of program testing. www.cl.cam.ac.uk/~carh4/19.Jan.18.Lecture1.pdf. Accessed 11 July 2019
36. Hoare, T., Mendes, A., Ferreira, J.F.: Logic, algebra, and geometry at the foundation of computer science. In: Formal Methods Teaching Workshop and Tutorial, FMTea 2019 (2019)
37. Hoare, T., Möller, B., Struth, G., Wehrman, I.: Concurrent Kleene algebra and its foundations. J. Log. Algebr. Program. **80**(6), 266–296 (2011)
38. Hoare, T., O'Hearn, P.W.: Separation logic semantics for communicating processes. Electr. Notes Theoret. Comput. Sci. **212**, 3–25 (2008)
39. Hoare, T., van Staden, S., Möller, B., Struth, G., Zhu, H.: Developments in concurrent Kleene algebra. J. Log. Algebr. Methods Program. **85**(4), 617–636 (2016)
40. Horn, A.: On sentences which are true of direct unions of algebras. J. Symb. Log. **16**(1), 14–21 (1951)
41. Ip, C.N., Dill, D.L.: Better verification through symmetry. Formal Methods Syst. Des. **9**(1/2), 41–75 (1996)
42. Jackson, D.: Alloy: a lightweight object modelling notation. ACM Trans. Softw. Eng. Methodol. **11**(2), 256–290 (2002)
43. Jervis, C. (ed.): ITU-T: Recommendation Z.120 (04/04), Message Sequence Charts (MSC). International Telecommunication Union, Geneva (2004)
44. Kolovos, D.S., Paige, R.F., Polack, F.A.C.: The epsilon transformation language. In: Vallecillo, A., Gray, J., Pierantonio, A. (eds.) ICMT 2008. LNCS, vol. 5063, pp. 46–60. Springer, Heidelberg (2008). https://doi.org/10.1007/978-3-540-69927-9_4
45. Kozen, D.: Kleene algebra with tests. ACM Trans. Program. Lang. Syst. **19**(3), 427–443 (1997)
46. Kwiatkowska, M., Norman, G., Parker, D.: PRISM 4.0: verification of probabilistic real-time systems. In: Gopalakrishnan, G., Qadeer, S. (eds.) CAV 2011. LNCS, vol. 6806, pp. 585–591. Springer, Heidelberg (2011). https://doi.org/10.1007/978-3-642-22110-1_47
47. Laurence, M.R., Struth, G.: Completeness theorems for pomset languages and concurrent Kleene algebras. CoRR abs/1705.05896 (2017)
48. MacNeille, H.M.: Partially ordered sets. Trans. AMS **42**(3), 416–460 (1937)
49. McIver, A., Rabehaja, T.M., Struth, G.: On probabilistic Kleene algebras, automata and simulations. In: de Swart, H. (ed.) RAMICS 2011. LNCS, vol. 6663, pp. 264–279. Springer, Heidelberg (2011). https://doi.org/10.1007/978-3-642-21070-9_20

50. McIver, A., Rabehaja, T.M., Struth, G.: Probabilistic concurrent Kleene algebra. In: Bortolussi, L., Wiklicky, H. (eds.) 11th International Workshop on Quantitative Aspects of Programming Languages and Systems, QAPL 2013, Rome, 23–24 March 2013. EPTCS, vol. 117, pp. 97–115 (2013)
51. Microsoft: Time Travel Debugging in WinDbg Preview! blogs.msdn.microsoft.com/windbg/2017/09/25/time-travel-debugging-in-windbg-preview/. Accessed 01 July 2019
52. Milner, R. (ed.): A Calculus of Communicating Systems. LNCS, vol. 92. Springer, Heidelberg (1980). https://doi.org/10.1007/3-540-10235-3
53. Miyazawa, A., Ribeiro, P., Li, W., Cavalcanti, A.L.C., Timmis, J., Woodcock, J.C.P.: RoboChart: modelling and verification of the functional behaviour of robotic applications. Softw. Syst. Model. (2019)
54. Möller, B., Hoare, T., Müller, M.E., Struth, G.: A discrete geometric model of concurrent program execution. In: Bowen, J.P., Zhu, H. (eds.) UTP 2016. LNCS, vol. 10134, pp. 1–25. Springer, Cham (2017). https://doi.org/10.1007/978-3-319-52228-9_1
55. de Moura, L., Kong, S., Avigad, J., van Doorn, F., von Raumer, J.: The Lean theorem prover (system description). In: Felty, A.P., Middeldorp, A. (eds.) CADE 2015. LNCS, vol. 9195, pp. 378–388. Springer, Cham (2015). https://doi.org/10.1007/978-3-319-21401-6_26
56. Huerta y Munive, J.J., Struth, G.: Verifying hybrid systems with modal Kleene algebra. In: Desharnais, J., Guttmann, W., Joosten, S. (eds.) RAMiCS 2018. LNCS, vol. 11194, pp. 225–243. Springer, Cham (2018). https://doi.org/10.1007/978-3-030-02149-8_14
57. Naumann, D.A., Sampaio, A., Silva, L.: Refactoring and representation independence for class hierarchies. Theoret. Comput. Sci. **433**, 60–97 (2012)
58. Nipkow, T., Wenzel, M., Paulson, L.C. (eds.): Isabelle/HOL—A Proof Assistant for Higher-Order Logic. LNCS, vol. 2283. Springer, Heidelberg (2002). https://doi.org/10.1007/3-540-45949-9
59. Object Management Group: OMG: Unified Modeling Language: Superstructure 2.0 (2003)
60. O'Hearn, P.W.: Separation logic. Commun. ACM **62**(2), 86–95 (2019)
61. Oliveira, M., Cavalcanti, A., Woodcock, J.: A UTP semantics for Circus. Formal Asp. Comput. **21**(1–2), 3–32 (2009)
62. Peterson, J.L.: Petri nets. ACM Comput. Surv. **9**(3), 223–252 (1977)
63. Petri, C.A.: Communication with automata. DTIC Res. Rep. AD0630125, Defense Tech. Inf. Cntr., Fort Belvoir, VA (1966)
64. Pratt, V.: Action logic and pure induction. In: van Eijck, J. (ed.) JELIA 1990. LNCS, vol. 478, pp. 97–120. Springer, Heidelberg (1991). https://doi.org/10.1007/BFb0018436
65. Ramos, R., Sampaio, A., Mota, A.: A semantics for UML-RT active classes via mapping into Circus. In: Steffen, M., Zavattaro, G. (eds.) FMOODS 2005. LNCS, vol. 3535, pp. 99–114. Springer, Heidelberg (2005). https://doi.org/10.1007/11494881_7
66. Reynolds, J.C.: Separation logic: a logic for shared mutable data structures. In: 17th IEEE Symposium on Logic in Computer Science (LICS 2002), Copenhagen, 22–25 July 2002, pp. 55–74. IEEE Computer Society (2002)
67. RoboTool: Graphical modelling, validation, and automatic generation of mathematical definitions for proof for RoboChart models. www.cs.york.ac.uk/robostar/robotool/

68. Sakallah, K.A.: Symmetry and satisfiability. In: Biere, A., Heule, M., van Maaren, H., Walsh, T. (eds.) Handbook of Satisfiability, Frontiers in Artificial Intelligence and Applications, vol. 185, pp. 289–338. IOS Press (2009)
69. Santos, T., Cavalcanti, A., Sampaio, A.: Object-orientation in the UTP. In: Dunne, S., Stoddart, B. (eds.) UTP 2006. LNCS, vol. 4010, pp. 18–37. Springer, Heidelberg (2006). https://doi.org/10.1007/11768173_2
70. Silva, L., Sampaio, A., Liu, Z.: Laws of object orientation with reference semantics. In: Cerone, A., Gruner, S. (eds.) 6th IEEE International Conference on Software Engineering and Formal Methods, SEFM 2008, Cape Town, 10–14 November 2008, pp. 217–226. IEEE Computer Society (2008)
71. Steinberg, D., Budinsky, F., Paternostro, M., Merks, E.: EMF: Eclipse Modeling Framework 2.0, 2nd edn. Addison-Wesley, Boston (2009)
72. Tarlecki, A.: A language of specified programs. Sci. Comput. Program. 5(1), 59–81 (1985)
73. Tarski, A.: A lattice-theoretical fixpoint theorem and its applications. Pac. J. Math. 5, 285–309 (1955)
74. Wing, J.M., Woodcock, J., Davies, J. (eds.): FM 1999. LNCS, vol. 1708. Springer, Heidelberg (1999). https://doi.org/10.1007/3-540-48119-2
75. Wing, J.M., Woodcock, J., Davies, J. (eds.): FM 1999. LNCS, vol. 1709. Springer, Heidelberg (1999). https://doi.org/10.1007/3-540-48118-4
76. Woodcock, J., Cavalcanti, A.: The semantics of *Circus*. In: Bert, D., Bowen, J.P., Henson, M.C., Robinson, K. (eds.) ZB 2002. LNCS, vol. 2272, pp. 184–203. Springer, Heidelberg (2002). https://doi.org/10.1007/3-540-45648-1_10
77. Woodcock, S.F.J., Zeyda, F.: Unifying semantic foundations for automated verification tools in Isabelle/UTP. CoRR abs/1905.05500 (2019)
78. von Wright, J.: Towards a refinement algebra. Sci. Comput. Program. 51(1–2), 23–45 (2004)
79. Ye, K., Woodcock, J., Foster, S., Miyazawa, A., Cavalcanti, A.: RoboChart: formal modelling and verification of the probabilistic behaviour of robotic applications. Technical report. University of York (2019)
80. Zeyda, F., Santos, T., Cavalcanti, A., Sampaio, A.: A modular theory of object orientation in higher-order UTP. In: Jones, C., Pihlajasaari, P., Sun, J. (eds.) FM 2014. LNCS, vol. 8442, pp. 627–642. Springer, Cham (2014). https://doi.org/10. 1007/978-3-319-06410-9_42

A Testing Perspective on Algebraic, Denotational, and Operational Semantics

Bernhard K. Aichernig[✉]

Institute of Software Technology, Graz University of Technology, Graz, Austria
`aichernig@ist.tugraz.at`

Abstract. In this paper, we discuss the role of formal semantics from a testing perspective. Our focus is on conformance testing, where we test if a given system-under-test conforms to an abstract description of its intended behaviour. We show how the main semantic paradigms, namely algebraic, denotational, and operational semantics, support a systematic testing process and give examples from our own work on automated test-case generation.

Keywords: Semantics · Unifying Theories of Programming · Test-case generation · Model-based testing · SMT solving · Symbolic execution · Quickcheck

1 Introduction

Is testing able to show the absence of bugs? The most prominent negative answer was given by the late Edsger Dijkstra: "Program testing can be a very effective way to show the presence of bugs, but it is hopelessly inadequate for showing their absence." [43]. Dijkstra was always motivating the need for formally verified software. Of course, in general Dijkstra is right, in the same way as Popper was right, when he stated that we can never verify that a theory is correct by a finite set of experiments. In principle, only refutation (falsification) is possible [64].However, this should not lead to an over-pessimistic judgement rejecting testing completely. This would be futile, since testing is the only way of building trust in a *running* system embedded in a complex environment. Testing is needed to check our assumptions. With wrong assumptions, even formally verified software may fail.

A famous example of such a rare and subtle software bug was found in the binary search algorithm implemented in the Java JDK 1.5 library in 2006 [38]. For large arrays the binary search method threw an exception raised by accessing the array out of its boundaries. The fault was in the line responsible for calculating the next element in the search following the divide-and-conquer strategy: `int mid = (lo + hi) / 2;`. From an algorithmic point of view this line is perfectly fine, assuming idealised infinite integers. However, this assumption is wrong in the case of a concrete computer with bounded integer ranges. For large `lo` and `hi`

© Springer Nature Switzerland AG 2019
P. Ribeiro and A. Sampaio (Eds.): UTP 2019, LNCS 11885, pp. 22–38, 2019.
https://doi.org/10.1007/978-3-030-31038-7_2

values the sum would create an overflow leading to a negative value of variable
`mid`. The developer of this method reported that he actually took the algorithm
including this line from the famous book Programming Pearls [36]. The fault
resided in Sun's Java library for nine years, before it was found. This code line
existed for two decades in the algorithm book without anybody noticing the
problem. The algorithm was even proved correct. So what was the nature of
this problem? The answer is that the fault was introduced by assuming a wrong
background theory on numbers. In the domain of mathematics the algorithm
works perfectly fine, assuming infinite integers. In the domain of Java with integer overflows it is wrong. The correctness proof relied on the wrong assumption
and therefore could not detect the problem.

This example shows that we have to keep in mind that program proof is
about proving a formula, model checking is about checking a model, but only
testing is targeting the running system in its real environment[1].

Furthermore, the example shows the essential role of semantics (here, the
interpretation of integer values).

Gaudel showed that testing can be formal too [45], and even one of the most
prominent figures in computer science, Tony Hoare, has changed his view:

> "I have radically changed my attitude towards program testing which I
> now understand to be entirely complementary to scientific design and ver-
> ification methods, and makes an equal contribution to the development of
> reliable software on industrial scale." [50]

In this paper, we go one step further and claim that **systematic testing
is able to show the absence of specific faults—under certain (strong)
assumptions.** In order to achieve this, we need precise fault models and systematic test-case generation methods. We argue that formal semantics is the
foundation for such advanced testing techniques. In the following, we will show
that the different semantic paradigms as presented in the Unifying Theories of
Programming (UTP) by Hoare and He [49] can support different testing processes and give examples from our own work.

We will limit ourselves to conformance testing with the goal of determining if
a program or, generally, a system-under-test (SUT), complies with the requirements of a specification, technical standard, or contract. In order to automate
conformance testing, we need a formal language to express these requirements.
In order to be useful, such a specification/modelling language must

– be expressive enough in order to represent the requirements in a given domain,
– support abstraction in order to specify/model *what* a SUT is supposed to
  compute, in contrast to *how* it computes, and
– be defined in a precise and unambigous semantics.

Structure. After covering, the necessary preliminaries in Sect. 2, in Sect. 3
we are going to discuss the role of a operational semantics in testing. Then, in

---

[1] Note that even software model checking relies on specialised (non-standard) run-time
environments that may behave differently to the deployed system.

Sect. 4 we show how a denotational semantics supports test-case generation with SMT solving. To complete the picture, we discuss property-based testing based on algebraic semantics in Sect. 5. Finally, we draw our conclusions in Sect. 6.

2 Preliminaries

2.1 Model-Based Testing

A prominent form of conformance testing is model-based testing. Model-based testing is a black-box testing technique focusing on the external behaviour of a SUT. Hence, we assume that we have no access to the internals of the SUT, like e.g., the source code. The test stimuli are automatically generated from an abstract model of the SUT. This test model is usually derived from the requirements. The model serves also as a test oracle providing the verdict (pass or fail) of a test-case execution. The models are expressed in special modelling languages that support the abstract specification of the central properties to be tested. A detailed introduction to model-based testing can be found in [72,73].

It would be futile to expect that one modelling language serves all needs. Several factors influence the choice of the modelling language, including the application domain, expressitivity and tool support. For example, we have applied the following modelling languages in model-based testing:

1. VDM [56] for testing air-traffic communication systems [13,51,52],
2. RAISE [48] for testing data type implementations [12,42],
3. OCL [62] for testing against UML contracts [24],
4. LOTOS [53] for testing communication protocols [11,23,30,31,74],
5. NuSMV language [40] for testing automotive controllers [44],
6. Creol [55] for testing distributed object-oriented systems [14,15,46,47,66, 67],
7. Spec# [35] for testing C# programs [57],
8. Qualitative Reasoning models [59] for testing continuous systems [9],
9. Extended Action Systems [34] for testing hybrid systems [5,8,39],
10. REO [33] for testing coordinated networks of components [3,60],
11. Symbolic Labelled Transition Systems [65] for testing communication protocols [54] and embedded systems [69],
12. Timed Automata [32] for testing real-time systems [19], and finally,
13. UML state-machine diagrams for testing embedded systems [7,58].

Note that if a modelling language has no formal semantics, we need to define one, e.g., by translation to a formal specification language, like for UML [58].

2.2 Model-Based Mutation Testing

Model-based mutation testing [2] is a special form of model-based testing where (1) a fault is injected into a model (mutation), then (2) a test-case that triggers this fault is generated from the model, and finally, (3) this test case is executed on the SUT in order to test if the modelled fault is hidden in the SUT. This process is repeated for all faults of interest.

Figure 1 summarises the process of model-based mutation testing. Like in classical model-based testing, the user creates a test model out of the given requirements. A test-case generator then analyses the model and generates an abstract test case (or a test suite). This test case is on the same abstraction level as the test model and includes expected outputs. A test driver maps the abstract test case to the concrete test interface of the SUT and executes the test case. The test driver compares the expected outputs with the actual outputs of the SUT and issues a verdict (pass or fail).

If the SUT conforms to the model, i.e. the SUT implements the model correctly, the verdict will always be *pass* (assuming that the tool chain generates sound test cases). In case of non-conformance ($\neg$ conforms), i.e. a bug exists, we may issue a *fail* verdict. However, due to the incompleteness of testing, we may miss the bug and issue a *pass* verdict. Dijkstra was referring to these incompleteness of testing when he pointed out that testing cannot show the absence of bugs. However, in model-based mutation testing, we can improve this situation considerably.

In model-based mutation testing, we mutate the models automatically and then generate an abstract test case that will cover this mutation. What this coverage means will be defined later, when we define the conformance relation. For now we want to point out an important difference to other testing techniques: if a bug exists and this bug is represented by the generated mutant, then the test case will find this bug. This important property is illustrated in Fig. 1 by the two conformance arrows: if the SUT does not conform to the model, but conforms to the mutant, the execution of the generated test case will result in a *fail* verdict. Here we are assuming a deterministic implementation. For non-deterministic SUTs, we have to assume a certain level of fairness and repeat the test cases a given number of times.

The notion of conformance defines when an injected fault leads to an observable failure. Hence, in order to reason about faults in testing, we need to define the conformance relation precisely. This is only possible when we fix the semantics of the modelling language. Note, the term *conformance relation* in testing is the equivalent of an *implementation relation* in formal methods.

In the following, we will define conformance in terms of operational, denotation and algebraic semantics and discuss how the different semantical styles support test-case generation.

3   Operational Semanctics

3.1   Overview

Operational semantics is very popular in concurrency theory and model checking. An operational semantics defines the meaning of a programming or modelling language in terms of abstract machines. It is operational in the sense that it explains the operational (execution) behaviour of a language. This is the oldest

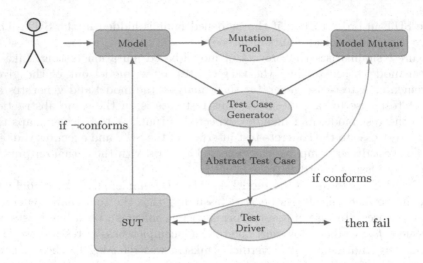

Fig. 1. Model-based mutation testing.

form of semantics and has been applied in compiler design since the 1960s. The virtual stack machine of Java is a prominent example of an operational semantics assigned to a programming language. However, for theory building, usually more abstract machines, represented as (labelled) transition systems, are used. Model checking is performed on various kinds of automata. For reasoning purposes, Plotkin has established the style of structured operational semantics [63]. Here the behaviour of these transition systems is presented in the form of formal proof rules over the abstract syntax of the language.

3.2 Conformance of Input-Output LTSs

A prominent testing theory for labelled transition systems with input and output labels was developed by Tretmans [71]. Its conformance relation *ioco* is defined as follows.

Definition 1.

$$SUT \text{ } ioco \text{ } Model =_{df} \forall \sigma \in traces(Model) : out(SUT \text{ } after \sigma) \subseteq out(Model \text{ } after \sigma)$$

*Here **after** denotes the set of reachable states after a trace σ, and **out** denotes the set of all observable events in a set of states. The observable events are all output events plus one additional quiescence event for indicating the absence of any output.*

This input-output conformance relation *ioco* supports non-deterministic models (see the subset relation) as well as partial models (only traces of the Model are tested). For input-complete models *ioco* is equivalent to trace-inclusion (language inclusion).

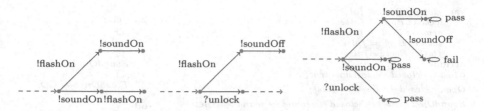

Fig. 2. Labelled transition systems of a part of a non-deterministic model of the car alarm system (left), a mutant (centre), and their synchronous product graph (right).

Example 1. Let us consider a car alarm system with sound and flash alarms. The left-hand side of Fig. 2 shows the LTS semantics of switching on both kind of alarms non-deterministically. Exclamation marks denote observable (output) events and question marks controllable (input) events. In this model either the flash, or the sound is switched on first. An implementer may decide for one of the two interleavings according to *ioco*. He might even add additional controllable events at any point, like the *?unlock*-event in the LTS at the centre. However, the subset relation of output events has to be respected. Therefore, it is the *!soundOff* event in the mutant in the centre causing non-conformance.

3.3 Explicit Conformance Checking

The conformance between a model and its mutant can be checked by building the synchronous product of their LTSs modulo *ioco*. The right-hand side of Fig. 2 shows this product graph for our example. Product modulo *ioco* means that we limit the standard product construction if the mutant has either (1) an additional (unexpected) output event (here *!soundOff*), or (2) an additional input event (here *?unlock*). In the first case, we have detected non-conformance and add a fail state after the unexpected event. Furthermore, we add all expected observables of the model. In the second case we stop the exploration, because we have reached an unspecified input behaviour.

Different strategies for extracting a test case from such a product graph exist. We can select a linear or adaptive test case, the shortest path or a random path to a fail-state, cover each fail-state or only one. Our experiments have shown that a combination of random and lazy shortest path strategies works well [6]. Lazy refers to the strategy of generating new test cases only, if the existing test cases do not kill a mutant.

We have applied this explicit conformance checking technique to generate test cases to several case studies, including testing an HTTP server [11] and SIP servers [74] using LOTOS models, controllers [6] using UML models, and most challenging, hybrid systems [39] using Back's Action Systems extended with qualitative reasoning models [9]. Our mapping from UML to an Action System is an operational interpretation of UML state machines in terms of Action Systems, too [6,58]. Let us have a closer look at Action Systems.

```
var  closed : Bool := false;                        . . .
     locked : Bool := false;                    do Close
     armed : Bool := false;                        □
     sound : Bool := false;                        Open
     flash : Bool := false;                        □
actions                                            SoundOn; FlashOn
Close :: ¬closed → closed := true;                 □
                                                   FlashOn; SoundOn
Open :: closed → closed := false;                  . . .
SoundOn :: armed ∧ ¬closed ∧ ¬sound → sound := true;  od
FlashOn :: armed ∧ ¬closed ∧ ¬flash → flash := true;
. . .
```

Fig. 3. Action System model of the car alarm system.

3.4 Action Systems

Back's Action Systems [34] are a kind of guarded command language for modelling concurrent reactive systems. They are similar to Dijkstra's iterative statement. Originally, their semantics is defined with weakest preconditions. However, for test-case generation, we have labelled the actions with input and output labels and interpret them as input-output labelled transitions systems.

Example 2. Figure. 3 shows an Action System model of our car alarm system example. First, the model variables and their initial values are declared. Next, the actions in the form of guarded commands are listed. Note that each action is labelled establishing the link to the LTS semantics. On the right-hand side is the protocol layer of actions which further restricts the possible order of actions. The standard composition operator for actions is non-deterministic choice ($A \square B$), however also sequential ($A; B$) or prioritised compositions ($A//B$) are possible. The protocol layer establishes a loop that iterates while any action is enabled. Action Systems terminate if all actions are disabled.   □

For explicit conformance checking we can explore the state space of the Action systems and construct the product graph of their LTS semantics. This works well for event-oriented systems, like protocols or embedded controllers. However, explicit checking does not scale well with more data-oriented models with parameterized events. These kind of models can be explored symbolically using symbolic execution.

3.5 Symbolic Conformance Checking

We have implemented a symbolic conformance checker that checks symbolic input-output conformance between a model and a mutant [28]. The algorithm is similar to the explicit conformance checking, but the exploration is realized via symbolic execution. More concretely, it performs a bounded depth-first search for states in which non-conformance may be observed. For this purpose, both mutant and specification are symbolically executed in parallel, such that they synchronise on observable actions, but execute internal actions independently from each other. At each step, a conformance check is performed and if non-conformance is detected, the trace leading to the current state and a non-conformance condition are returned.

In practice, several optimisations have been implemented: (1) If during search we reach a symbolic state which has already been visited, we prune the search tree. (2) We precompute symbolic execution graphs, which encode all executable traces for the specification model. This information can be reused during the conformance check and results in a performance gain, as we check conformance for hundreds of different mutants with the same specification model. (3) As long as the syntactic mutation has not been executed, the precomputed execution graph of the specification model can be used for the mutant as well.

We have also shown that it is possible to extend this approach to real-time systems [20]. In this work, we extended action systems with clock variables and checked a timed version of input-output conformance.

4   Denotational Semantics

A denotational semantics is defined via a mapping from syntax to a semantic domain. For example in case of imperative languages, this interpretation function maps every statement to a theory over the observations before and after its execution. Here, only the effects of what is computed are defined and not, like in operational semantics, how this computation is actually realised.

In UTP [49], Hoare and He propose a predicative semantics that can be exploited for test-case generation with constraint, SAT or SMT solvers.

4.1   Transformational Systems

Transformational systems transform inputs and a pre-state to some output and post-state, then they terminate. Hence, the model and mutant of a transformational system can be interpreted as predicates $Model(s, s')$ and $Mutant(s, s')$ describing their state transformations $(s \rightarrow s')$. For such relational models, conformance is defined via implication in the standard way:

Definition 2 (Conformance as Implication).

$$Model \sqsubseteq Mutant =_{df} \forall s, s' : Mutant(s, s') \Rightarrow Model(s, s')$$

Here conformance between a mutant and a model means that all behaviour of the mutant is allowed by the model. Consequently, non-conformance is expressed via the existence of a behaviour of the mutant that is not allowed by the model:

Theorem 1.

$$Model \not\sqsubseteq Mutant = \exists s, s' : Mutant(s, s') \land \neg Model(s, s')$$

The above formula represents a constraint-satisfaction problem. Hence, a constraint solver or SMT solver can be used to search for an input (pre-state) s leading to the fault.

$$l :: g \rightarrow B \qquad =_{df} g \ \wedge \ B \ \wedge \ tr' = tr \ \widehat{\ } \ [l]$$
$$l(\overline{x}) :: g \rightarrow B \qquad =_{df} \exists \ \overline{x} : g \ \wedge \ B \ \wedge \ tr' = tr \ \widehat{\ } \ [l(\overline{x})]$$
$$x := e \qquad =_{df} x' = e \ \wedge \ y' = y \ \wedge \dots \wedge \ z' = z$$
$$g \rightarrow B \qquad =_{df} g \ \wedge \ B$$
$$B(s, s'); B(s, s') =_{df} \exists \ s_0 : B(s, s_0) \ \wedge \ B(s_0, s')$$
$$B_1 \ \square \ B_2 \qquad =_{df} B_1 \ \vee \ B_2$$

Fig. 4. Predicative UTP semantics of Action System.

In our early work on OCL contracts [24], we gave OCL such a denotational semantics and developed a set-based constraint solver for test-case generation. Similarly, the work on Spec# [57] and REO [60] is based on this kind of denotational semantics of the specification languages.

We generalised this approach in our testing theory of mutation testing, where we gave test cases a denotational semantics and related them via refinement to models and implementations [1,16]. This enabled us to formally prove our mutation-based test-case generation algorithms being correct.

4.2   Action Systems

Test-case generation with predicative semantics can also be applied to Action Systems. We have given Action Systems a predicative semantics in the style of UTP as shown in Fig. 4.

The state-changes of actions are defined via predicates relating the pre-state of variables s and their post-state s'. Furthermore, the labels form a visible trace of events tr that is updated to tr' whenever an action runs through. Hence, a guarded action's transition relation is defined as the conjunction of its guard g, the body of the action B and the adding of the action label l to the previously observed trace. In case of parameters $\overline{x}$, these are added as local variables to the predicate. An assignment updates one variable x with the value of an expression e and leaves the rest unchanged. Sequential composition is standard: there must exist an intermediate state s_0 that can be reached from the first body predicate and from which the second body predicate can lead to its final state. Finally, non-deterministic choice is defined as disjunction[2].

Note that the existential quantifiers in the semantic definitions need to be eliminated before we can apply our formula for test-case generation. The negation of the original model would turn this into a universally quantified predicate which is beyond constraint solving. SMT solvers can handle universal quantification, however, our evaluation with Z3, showed low performance and that it is better to eliminate the existential quantifiers beforehand [17].

---

[2] In contrast to designs in UTP, we assume termination and, therefore, avoid the introduction of auxiliary ok variables. This is possible, since for test-case generation the outer loop is unfolded a finite number of times.

Action Systems are not transformational, but reactive systems. We have to take the looping semantics into account. Therefore, we have to add a reachability check to the non-conformance check. The reachability analysis is based on the bounded model-checking algorithm proposed by Biere et al. [37]. The obtained trace can be used to test, whether a system implements the original specification or a mutant.

Definition 3 (Fault-Based Test-Case Generation). *Given the transition relations T_{spec} and T_{mutant} and a bound k, the formula*

$$T_{mutant}(u, p) \wedge \neg T_{spec}(u, p) \wedge I(s_0) \wedge$$

$$\bigwedge_{i=0}^{k-1} T_{spec}(s_i, s_{i+1}) \wedge \bigvee_{i=0}^{k} s_i = u$$

is satisfiable iff there exists a trace leading to a faulty behaviour. A model of this formula contains this trace.

The action system is translated to an SMT formula representing the step relation T according to the semantics given in Fig. 4. We want to find an unsafe state u where the transition relation T_{mutant} does not conform to the transition relation T_{spec}. The formula $T_{mutant}(u, p) \wedge \neg T_{spec}(u, p)$ finds a state u where a non conforming step is possible. It is unsatisfiable if the mutant conforms to the specification. In this case it is not possible to generate a test case. The rest of the formula states as symbolic model-checking problem whether an unsafe state is reachable within k steps. The state variables in s_0 are set equal to the values assigned during initialisation of the Action System and the transition relation is unrolled k times: $I(s_0) \wedge T_{spec}(s_0, s_1) \wedge T_{spec}(s_1, s_2) \wedge \cdots \wedge T_{spec}(s_{k-1}, s_k)$. Then it is checked if one of the states is a non-conforming state: $s_0 = u \vee s_1 = u \vee \cdots \vee s_k = u$.

This technique was used to generate mutation test-cases from Action Systems with the SMT solver Z3 [21]. We also applied this bounded conformance-checking approach to deterministic Timed Automata [19], giving Timed Automata a predicative semantics.

5   Algebraic Semantics

Gaudel defined her testing theory [45] in this semantic style. An algebraic semantics defines the meaning of syntax by enumerating its algebraic properties. The properties are given in the form of equational axioms, possibly with preconditions. Therefore, this style is also called axiomatic semantics. Boolean algebra is an example of an algebraic semantics for Boolean expressions or circuits. The style has been successfully applied to the definition of abstract data types (ADTs).

Claessen and Hughes [41] showed with their tool QuickCheck how algebraic properties can be exploited for the automated testing of programs. The idea is best explained with a simple example: Consider the following algebraic property that the reverse of the reverse of a list must be equal to the original list:

$$\forall xs \in List[T] : reverse(reverse(xs)) = xs$$

QuickCheck generates a series of random lists xs and evaluates the equation with calls to the function $reverse$. Hence, QuickCheck generates random values according to the types of the universally quantified variables. This is realised via type-dependent generators that can be nested in order to generate test-data for nested types. For the list in our example, the list generator would invoke a generator for the type T in order to generate the inner elements of the list. In addition to the default generators for the standard types, the programmer can define custom generators.

In case the property is violated, QuickCheck reports the failing test case. Since random tests tend to be rather complicated, QuickCheck tries to find a simpler failing test-cases for debugging. This is known as Shrinking and the user is able to specify the shrinking policy individually for different data types.

Due to its flexibility and scalability, this testing technique became very popular among programmers and is known as Property-Based Testing. Today, there are many other tools that are based on the concepts of QuickCheck, e.g., ScalaCheck [61] for Scala, Hypothesis[3] for Python, and FsCheck[4] for C# and F#.

In our own work, we have recently applied property-based testing for the load testing of an industrial web application [27] and different open-source MQTT brokers [26]. MQTT is a publish-subscribe IoT protocol. By combining property-based testing with statistical model checking algorithms [25], we could estimate the expected latencies for different user [68] and deployment scenarios [18].

6   Concluding Remarks

In this paper we wanted to highlight the relation between testing and semantics. Testing is about finding bugs and in order to define what behaviour constitutes a bug, we need semantics. Only with precise semantics it is possible to define conformance and non-conformance. Furthermore, we showed how the different semantic paradigms relate to test-case generation.

In operational semantics, a specification model is interpreted as an abstract machine. This machine is then explored according to its operational semantics, concretely or symbolically, in order to find an adequate test case. In model-based mutation testing, the aim is to find a test-case that can distinguish the model from the mutated model. This test goal is defined as non-conformance.

In denotational semantics, we interpret a specification model as a predicate describing its possible behaviour. We can then use a constraint or SMT solver to generate test-cases. For model-based mutation testing, we apply bounded conformance checking.

---

[3] https://pypi.python.org/pypi/hypothesis.
[4] https://github.com/fscheck/FsCheck.

In algebraic semantics, an abstract property is used to test the system. Universal quantifiers are tested with random values generated from type-dependent generators. This property-based testing style can be combined with statistical testing methods in order to obtain stochastic guarantees.

As argued, precise semantics is the basis for any formal testing process that is based on sound theory. Within such a testing theory we can state our assumptions clearly and formulate correctness guarantees provided these assumptions hold. Within the discussed model-based mutation testing framework, we anticipate a set of specific faults and then generate test-cases that guarantee to detect these specific faults. We have proved this in UTP via refinement [16]. Obviously, we rely on several assumptions: (1) the SUT behaves according to our semantics, (2) the faults we anticipate are representative, (3) either the system is deterministic or we have some statistical measure how often tests need to be performed. Hence, the guarantees obtained via testing are conditional, but this is better than the hopeless characterisation of Dijkstra. Once, these testing assumptions are explicit, we can argue, reason and experiment if they hold for a given class of systems.

We hope that we could convince the reader that serious testing needs an understanding of semantics. From semantics, we build our testing theories, from the testing theories we derive our algorithms and tools. This principle guides our own research on model-based test-case generation. We are convinced that this is not the end of the journey.

As systems become more complex, intelligent, and adaptive, we will need new testing methods that can cope with these challenges. Currently, we are working on learning-based testing methods that combine model-based testing and model learning [22]. The first results are encouraging [4, 10, 29, 70]. However, without a good understanding of semantics in our toolbox, we would be lost.

Acknowledgment. This work was supported by the TU Graz LEAD project "Dependable Internet of Things in Adverse Environments".

References

1. Aichernig, B.K.: Mutation testing in the renement calculus. Formal Aspects Comput. **15**(2–3), 280–295 (2003)
2. Aichernig, B.K.: Model-based mutation testing of reactive systems. In: Liu, Z., Woodcock, J., Zhu, H. (eds.) Theories of Programming and Formal Methods. LNCS, vol. 8051, pp. 23–36. Springer, Heidelberg (2013). https://doi.org/10.1007/978-3-642-39698-4_2
3. Aichernig, B.K., Arbab, F., Astefanoaei, L., de Boer, F.S., Sun, M., Rutten, J.: Fault-based test case generation for component connectors. In: TASE 2009, Third IEEE International Symposium on Theoretical Aspects of Software Engineering, Tianjin, China, 29–31 July , pp. 147–154. IEEE Computer Society, July 2009

4. Aichernig, B.K., Bloem, R., Ebrahimi, M., Tappler, M., Winter, J.: Automata learning for symbolic execution. In: 2018 Formal Methods in Computer Aided Design, FMCAD 2018, Austin, TX, USA, October 30 - November 2 2018, pp. 1–9. IEEE (2018)
5. Aichernig, B.K., Brandl, H., Jöbstl, E., Krenn, W.: Model-based mutation testing of hybrid systems. In: de Boer, F.S., Bonsangue, M.M., Hallerstede, S., Leuschel, M. (eds.) FMCO 2009. LNCS, vol. 6286, pp. 228–249. Springer, Heidelberg (2010). https://doi.org/10.1007/978-3-642-17071-3_12
6. Aichernig, B.K., Brandl, H., Jöbstl, E., Krenn, W.: Efficient mutation killers in action. In: IEEE Fourth International Conference on Software Testing, Verification and Validation, ICST 2011, Berlin, Germany, 21–25 March 2011, pp. 120–129. IEEE Computer Society (2011)
7. Aichernig, B.K., Brandl, H., Jöbstl, E., Krenn, W.: UML in action: a two-layered interpretation for testing. ACM SIGSOFT Softw. Eng. Notes 36(1), 1–8 (2011)
8. Aichernig, B.K., Brandl, H., Krenn, W.: Qualitative action systems. In: Breitman, K., Cavalcanti, A. (eds.) ICFEM 2009. LNCS, vol. 5885, pp. 206–225. Springer, Heidelberg (2009). https://doi.org/10.1007/978-3-642-10373-5_11
9. Aichernig, B.K., Brandl, H., Wotawa, F.: Conformance testing of hybrid systems with qualitative reasoning models. In: Finkbeiner, B., Gurevich, Y., Petrenko, A.K. (eds.) Proceedings of Fifth Workshop on Model Based Testing (MBT 2009), York, England, 22 March 2009, volume 253 (2) of Electronic Notes in Theoretical Computer Science, pp. 53–69. Elsevier, October 2009
10. Aichernig, B.K., Burghard, C., Korošec, R.: Learning-based testing of an industrial measurement device. In: Badger, J.M., Rozier, K.Y. (eds.) NFM 2019. LNCS, vol. 11460, pp. 1–18. Springer, Cham (2019). https://doi.org/10.1007/978-3-030-20652-9_1
11. Aichernig, B.K., Delgado, C.C.: From faults via test purposes to test cases: on the fault-based testing of concurrent systems. In: Baresi, L., Heckel, R. (eds.) FASE 2006. LNCS, vol. 3922, pp. 324–338. Springer, Heidelberg (2006). https://doi.org/10.1007/11693017_24
12. Aichernig, B.K., George, C.: When model-based testing fails. In: Finkbeiner, B., Gurevich, Y., Petrenko, A.K. (eds.) Proceedings of the Second Workshop on Model Based Testing (MBT 2006), Second Workshop on Model Based Testing 2006, volume 164 (4), Electronic Notes in Theoretical Computer Science, pp. 115–128. Elsevier (2006)
13. Aichernig, B.K., Gerstinger, A., Aster, R.: Formal specification techniques as a catalyst in validation. In: Proceedings of the 5th IEEE High Assurance Systems Engineering Symposium (HASE 2000), Albuquerque, New Mexico, 15–17 November, pp. 203–207. IEEE (2000)
14. Aichernig, B.K., Griesmayer, A., Johnsen, E.B., Schlatte, R., Stam, A.: Conformance testing of distributed concurrent systems with executable designs. In: de Boer, F.S., Bonsangue, M.M., Madelaine, E. (eds.) FMCO 2008. LNCS, vol. 5751, pp. 61–81. Springer, Heidelberg (2009). https://doi.org/10.1007/978-3-642-04167-9_4
15. Aichernig, B.K., Griesmayer, A., Schlatte, R., Stam, A.: Modeling and testing multi-threaded asynchronous systems with Creol. In: Proceedings of the 2nd International Workshop on Harnessing Theories for Tool Support in Software (TTSS 2008), Istanbul, Turkey, 30 August 2008, volume 243 of Electronic Notes in Theoretical Computer Science, pp. 3–14. Elsevier, July 2009
16. Aichernig, B.K., He, J.: Mutation testing in UTP. Formal Aspects Comput. 21(1–2), 33–64 (2009)

17. Aichernig, B.K., Jöbstl, E., Kegele, M.: Incremental refinement checking for test case generation. In: Veanes, M., Viganò, L. (eds.) TAP 2013. LNCS, vol. 7942, pp. 1–19. Springer, Heidelberg (2013). https://doi.org/10.1007/978-3-642-38916-0_1

18. Aichernig, B.K., Kann, S., Schumi, R.: Statistical model checking of response times for different system deployments. In: Feng, X., Müller-Olm, M., Yang, Z. (eds.) SETTA 2018. LNCS, vol. 10998, pp. 153–169. Springer, Cham (2018). https://doi.org/10.1007/978-3-319-99933-3_11

19. Aichernig, B.K., Lorber, F., Ničković, D.: Time for mutants—model-based mutation testing with timed automata. In: Veanes, M., Viganò, L. (eds.) TAP 2013. LNCS, vol. 7942, pp. 20–38. Springer, Heidelberg (2013). https://doi.org/10.1007/978-3-642-38916-0_2

20. Aichernig, B.K., Lorber, F., Tappler, M.: Conformance checking of real-time models. In: Ábrahám, E., Bonsangue, M., Johnsen, E.B. (eds.) Theory and Practice of Formal Methods. LNCS, vol. 9660, pp. 15–32. Springer, Cham (2016). https://doi.org/10.1007/978-3-319-30734-3_4

21. Aichernig, B.K., Maderbacher, B., Tiran, S.: Programming behavioral test models for SMT solving in Scala. In: 2019 IEEE International Conference on Software Testing, Verification and Validation Workshops, ICST Workshops 2019, Xi'an, China, 22–23 April 2019, pp. 52–60. IEEE (2019)

22. Aichernig, B.K., Mostowski, W., Mousavi, M.R., Tappler, M., Taromirad, M.: Model learning and model-based testing. In: Bennaceur, A., Hähnle, R., Meinke, K. (eds.) Machine Learning for Dynamic Software Analysis: Potentials and Limits. LNCS, vol. 11026, pp. 74–100. Springer, Cham (2018). https://doi.org/10.1007/978-3-319-96562-8_3

23. Aichernig, B.K., Peischl, B., Weiglhofer, M., Wotawa, F.: Protocol conformance testing a SIP registrar: an industrial application of formal methods. In: Hinchey, M., Margaria,T. (eds.) Fifth IEEE International Conference on Software Engineering and Formal Methods (SEFM 2007), London, England, UK, 10–14 September 2007, pp. 215–226. IEEE Computer Society, 2007

24. Aichernig, B.K., Salas, P.A.P.: Test case generation by OCL mutation and constraint solving. In: Cai, K.-Y., Ohnishi, A. (eds.) QSIC 2005, Fifth International Conference on Quality Software, Melbourne, Australia, 19–21 September 2005, pp. 64–71. IEEE Computer Society, 2005

25. Aichernig, B.K., Schumi, R.: Statistical model checking meets property-based testing. In: 2017 IEEE International Conference on Software Testing, Verification and Validation, ICST 2017, Tokyo, Japan, 13–17 March 2017, pp. 390–400. IEEE Computer Society, 2017

26. Aichernig, B.K., Schumi, R.: How fast is MQTT? In: McIver, A., Horvath, A. (eds.) QEST 2018. LNCS, vol. 11024, pp. 36–52. Springer, Cham (2018). https://doi.org/10.1007/978-3-319-99154-2_3

27. Aichernig, B.K., Schumi, R.: Property-based testing of web services by deriving properties from business-rule models. Softw. Syst. Model. 18(2), 889–911 (2019). Open Access

28. Aichernig, B.K.,Tappler, M.: Symbolic input-output conformance checking for model-based mutation testing. In: The 1st workshop on Uses of Symbolic Execution (USE), Oslo, Norway, 23–25 June 2015. Proceedings, Electronics Notes in Theoretical Computer Science, 320:3–19 (2016)

29. Aichernig, B.K., Tappler, M.: Probabilistic black-box reachability checking. In: Lahiri, S., Reger, G. (eds.) RV 2017. LNCS, vol. 10548, pp. 50–67. Springer, Cham (2017). https://doi.org/10.1007/978-3-319-67531-2_4

30. Aichernig, B.K., Eiglhofer, M., Peischl, B., Wotawa, F.: Test purpose generation in an industrial application. In: Proceedings of the 3rd Workshop on Advances in Model Based Testing, A-MOST 2007, co-located with the ISSTA 2007 International Symposium on Software Testing and Analysis, London, United Kingdom, 9–12 July, pp. 115–125 (2007)

31. Aichernig, B.K., Weiglhofer, M., Wotawa, F.: Improving fault-based conformance testing. In: Finkbeiner, B., Gurevich,Y., Petrenko, A.K. (eds.) Proceedings of the Fourth Workshop on Model Based Testing (MBT 2008), volume 220 (1) of Electronic Notes in Theoretical Computer Science, pp. 63–77. Elsevier (2008)

32. Alur, R., Dill, D.L.: A theory of timed automata. Theor. Comput. Sci. **126**(2), 183–235 (1994)

33. Arbab, F.: Reo: a channel-based coordination model for component composition. Math. Struct. Comput. Sci. **14**(3), 329–366 (2004)

34. Back, R.J., Kurki-Suonio, R.: Decentralization of process nets with centralized control. In: Proceedings of the 2nd ACM SIGACT-SIGOPS Symposium on Principles of Distributed Computing, Montreal, Quebec, Canada, pp. 131–142. ACM (1983)

35. Barnett, M., Leino, K.R.M., Schulte, W.: The spec# programming system: an overview. In: Barthe, G., Burdy, L., Huisman, M., Lanet, J.-L., Muntean, T. (eds.) CASSIS 2004. LNCS, vol. 3362, pp. 49–69. Springer, Heidelberg (2005). https://doi.org/10.1007/978-3-540-30569-9_3

36. Bentley, J.: Progamming Pearls, 2nd edn. Addison-Wesley, Boston (2000)

37. Biere, A., Cimatti, A., Clarke, E., Zhu, Y.: Symbolic model checking without BDDs. In: Cleaveland, W.R. (ed.) TACAS 1999. LNCS, vol. 1579, pp. 193–207. Springer, Heidelberg (1999). https://doi.org/10.1007/3-540-49059-0_14

38. Bloch, J.: Extra, extra - read all about it: Nearly all binary searches and mergesorts are broken. Google Research Blog, June 2006. http://googleresearch.blogspot.com/2006/06/extra-extra-read-all-about-it-nearly.html (Accessed 17 Aug 2019)

39. Brandl, H., Weiglhofer, M., Aichernig, B.K.: Automated conformance verification of hybrid systems. In: Wang, J., Chan, W.K., Kuo, F.C. (eds.) Proceedings of the 10th International Conference on Quality Software, QSIC 2010, Zhangjiajie, China, 14–15 July 2010, pp. 3–12. IEEE Computer Society (2010)

40. Cimatti, A., Clarke, E., Giunchiglia, F., Roveri, M.: NuSMV: a new symbolic model verifier. In: Halbwachs, N., Peled, D. (eds.) CAV 1999. LNCS, vol. 1633, pp. 495–499. Springer, Heidelberg (1999). https://doi.org/10.1007/3-540-48683-6_44

41. Claessen, K., Hughes, J.: QuickCheck: a lightweight tool for random testing of Haskell programs. In: Proceedings of the Fifth ACM SIGPLAN International Conference on Functional Programming (ICFP 2000), Montreal, Canada, 18–21 September 2000, pp. 268–279. ACM (2000)

42. Dan, L., Aichernig, B.K.: Combining algebraic and model-based test case generation. In: Liu, Z., Araki, K. (eds.) ICTAC 2004. LNCS, vol. 3407, pp. 250–264. Springer, Heidelberg (2005). https://doi.org/10.1007/978-3-540-31862-0_19

43. Dijkstra, E.W.: The humble programmer. Commun. ACM **15**(10), 859–866 (1972)

44. Fraser, G., Aichernig, B.K., Wotawa, F.: Handling model changes: regression testing and test-suite update with model-checkers. In: Proceedings of the Third Workshop on Model Based Testing (MBT 2007), volume 190(2) of Electronic Notes in Theoretical Computer Science, pp. 33–46. Elsevier (2007)

45. Gaudel, M.-C.: Testing can be formal, too. In: Mosses, P.D., Nielsen, M., Schwartzbach, M.I. (eds.) CAAP 1995. LNCS, vol. 915, pp. 82–96. Springer, Heidelberg (1995). https://doi.org/10.1007/3-540-59293-8_188

46. Grabe, I., et al.: Credo methodology: modeling and analyzing a peer-to-peer system in credo. Electron. Notes Theor. Comput. Sci. **266**, 33–48 (2010). Proceedings of the 3rd International Workshop on Harnessing Theories for Tool Support in Software (TTSS)

47. Griesmayer, A., Aichernig, B., Johnsen, E.B., Schlatte, R.: Dynamic symbolic execution for testing distributed objects. In: Dubois, C. (ed.) TAP 2009. LNCS, vol. 5668, pp. 105–120. Springer, Heidelberg (2009). https://doi.org/10.1007/978-3-642-02949-3_9

48. The RAISE Method Group: The RAISE Development Method. The BCS Practitioners Series. Prentice-Hall, Upper Saddle River (1995)

49. Hoare, C.A.R., He, J.: Unifying Theories of Programming. Prentice-Hall International, Upper Saddle River (1998)

50. Hoare, T.: Towards the verifying compiler. In: Aichernig, B.K., Maibaum, T. (eds.) Formal Methods at the Crossroads. From Panacea to Foundational Support. LNCS, vol. 2757, pp. 151–160. Springer, Heidelberg (2003). https://doi.org/10.1007/978-3-540-40007-3_10

51. Hörl, J., Aichernig, B.K.: Requirements validation of a voice communication system used in air traffic control, an industrial application of light-weight formal methods (abstract). In: Proceedings of the Fourth International Conference on Requirements Engineering (ICRE2000), Schaumburg, Illinois, 19–23 June 2000, pp. 190. IEEE (2000). Selected as one of three best papers

52. Hörl, J., Aichernig, B.K.: Validating voice communication requirements using lightweight formal methods. IEEE Softw. **17**(3), 21–27 (2000). **Best paper award** at Fourth International Conference on Requirements Engineering (ICRE2000)

53. ISO. ISO 8807: Information processing systems - open systems interconnection - LOTOS - a formal description technique based on the temporal ordering of observational behaviour (1989)

54. Jöbstl, E., Weiglhofer, M., Aichernig, B.K., Wotawa, F.: When BDDs fail: conformance testing with symbolic execution and SMT solving. In: Third International Conference on Software Testing, Verification and Validation (ICST 2010), Paris, France, 7–9 April 2010, pp. 479–488. IEEE Computer Society (2010)

55. Johnsen, E.B., Owe, O.: An asynchronous communication model for distributed concurrent objects. Softw. Syst. Model. **6**(1), 35–58 (2007)

56. Jones, C.B.: Systematic Software Development Using VDM. Series in Computer Science, 2nd edn. Prentice-Hall, Upper Saddle River (1990)

57. Krenn, W., Aichernig, B.K.: Test case generation by contract mutation in Spec#. In: Finkbeiner, B., Gurevich, Y., Petrenko, A.K. (eds.) Proceedings of Fifth Workshop on Model Based Testing (MBT 2009), York, England, 22 March 2009, volume 253 (2) of Electronic Notes in Theoretical Computer Science, pp. 71–86. Elsevier (2009)

58. Krenn, W., Schlick, R., Aichernig, B.K.: Mapping UML to labeled transition systems for test-case generation. In: de Boer, F.S., Bonsangue, M.M., Hallerstede, S., Leuschel, M. (eds.) FMCO 2009. LNCS, vol. 6286, pp. 186–207. Springer, Heidelberg (2010). https://doi.org/10.1007/978-3-642-17071-3_10

59. Kuipers, B.: Qualitative Reasoning: Modeling and Simulation with Incomplete Knowledge. MIT Press, Cambridge (1994)

60. Meng, S., Arbab, F., Aichernig, B.K., Astefanoaei, L., de Boer, F.S., Rutten, J.: Connectors as designs: modeling, refinement and test case generation. Sci. Comput. Program. **77**(7/8), 799–822 (2012)

61. Nilsson, R.: ScalaCheck: The Definitive Guide. IT Pro, Artima Incorporated, Walnut Creek (2014)

62. OMG. Object constraint language, version 2.2. Technical Report formal/2010-02-01, Object Management Group, February 2010
63. Plotkin, G.D.: A structural approach to operational semantics. Technical Report DAIMI FN-19, Computer Science Department, Aarhus University (1981)
64. Popper, K.: Logik der Forschung, 10th edn. Mohr Siebeck, Heidelberg (2005)
65. Rusu, V., du Bousquet, L., Jéron, T.: An approach to symbolic test generation. In: Grieskamp, W., Santen, T., Stoddart, B. (eds.) IFM 2000. LNCS, vol. 1945, pp. 338–357. Springer, Heidelberg (2000). https://doi.org/10.1007/3-540-40911-4_20
66. Schlatte, R., Aichernig, B., de Boer, F., Griesmayer, A., Johnsen, E.B.: Testing concurrent objects with application-specific schedulers. In: Fitzgerald, J.S., Haxthausen, A.E., Yenigun, H. (eds.) ICTAC 2008. LNCS, vol. 5160, pp. 319–333. Springer, Heidelberg (2008). https://doi.org/10.1007/978-3-540-85762-4_22
67. Schlatte, R., Aichernig, B., Griesmayer, A., Kyas, M.: Resource modeling for timed Creol models. Electron. Notes Theor. Comput. Sci. **266**, 63–75 (2010)
68. Schumi, R., Lang, P., Aichernig, B.K., Krenn, W., Schlick, R.: Checking response-time properties of web-service applications under stochastic user profiles. In: Yevtushenko, N., Cavalli, A.R., Yenigün, H. (eds.) ICTSS 2017. LNCS, vol. 10533, pp. 293–310. Springer, Cham (2017). https://doi.org/10.1007/978-3-319-67549-7_18
69. Schwarzl, C., Aichernig, B.K., Wotawa, F.: Compositional random testing using extended symbolic transition systems. In: Wolff, B., Zaïdi, F. (eds.) ICTSS 2011. LNCS, vol. 7019, pp. 179–194. Springer, Heidelberg (2011). https://doi.org/10.1007/978-3-642-24580-0_13
70. Tappler, M., Aichernig, B.K., Bloem, R.: Model-based testing IoT communication via active automata learning. In: 2017 IEEE International Conference on Software Testing, Verification and Validation, ICST 2017, Tokyo, Japan, 13–17 March 2017, pp. 276–287. IEEE Computer Society (2017)
71. Tretmans, J.: Test generation with inputs, outputs and repetitive quiescence. Softw. - Concepts Tools **17**(3), 103–120 (1996)
72. Utting, M., Legeard, B.: Practical Model-Based Testing: A Tools Approach. Morgan Kaufmann Publishers, Burlington (2007)
73. Utting, M., Pretschner, A., Legeard, B.: A taxonomy of model-based testing approaches. Softw. Test. Verif. Reliab. **22**(5), 297–312 (2011)
74. Weiglhofer, M., Aichernig, B.K., Wotawa, F.: Fault-based conformance testing in practice. Int. J. Softw. Inform. **3**(2–3), 375–411 (2009). Special double issue on Formal Methods of Program Development edited by Dines Bjoerner

Whither Specifications as Programs

David A. Naumann[✉] and Minh Ngo

Stevens Institute of Technology, Hoboken, USA
naumann@cs.stevens.edu, nngo1@stevens.edu

Abstract. Unifying theories distil common features of programming
languages and design methods by means of algebraic operators and
their laws. Several practical concerns—e.g., improvement of a program,
conformance of code with design, correctness with respect to specified requirements—are subsumed by the beautiful notion that programs
and designs are special forms of specification and their relationships
are instances of logical implication between specifications. Mathematical
development of this idea has been fruitful but limited to an impoverished
notion of specification: trace properties. Some mathematically precise
properties of programs, dubbed hyperproperties, refer to traces collec-
tively. For example, confidentiality involves knowledge of possible traces.
This article reports on both obvious and surprising results about lifting
algebras of programming to hyperproperties, especially in connection
with loops, and suggests directions for further research. The technical
results are: a compositional semantics, at the hyper level, of imperative
programs with loops, and proof that this semantics coincides with the
direct image of a standard semantics, for subset closed hyperproperties.

1 Introduction

A book has proper spelling provided that each of its sentences does. For a book
to be captivating and suspenseful—that is not a property that can be reduced
to a property of its individual sentences. Indeed, few interesting properties of a
book are simply a property of all its sentences. By contrast, many interesting
requirements of a program can be specified as so-called *trace properties*: there
is some property of traces (i.e., observable behaviors) which must be satisfied by
all the program's traces.

The unruly mess of contemporary programming languages, design tools, and
approaches to formal specification has been given a scientific basis through unifying theories that abstract commonalities by means of algebraic operators and
laws. Algebra abstracts from computational notions like partiality and nondeterminacy by means of operators that are interpreted as total functions and which
enable equational reasoning. Several practical concerns—such as improving a
program's resource usage while not altering its observable behavior, checking
conformance of code with design architecture, checking satisfaction of requirements, and equivalence of two differently presented designs—are subsumed by

© Springer Nature Switzerland AG 2019
P. Ribeiro and A. Sampaio (Eds.): UTP 2019, LNCS 11885, pp. 39–61, 2019.
https://doi.org/10.1007/978-3-030-31038-7_3

the beautiful notion that programs and designs[1] are just kinds of specification and their relationships are instances of logical implication between specifications. Transitivity of implication yields the primary relationship: the traces of a program are included in the traces allowed by its specification. The mathematical development of this idea has been very successful—for trace properties.

Not all requirements are trace properties. A program should be easy to read, consistent with dictates of style, and amenable to revision for adapting to changed requirements. Some though not all such requirements may be addressed by mathematics; e.g., parametric polymorphism is a form of modularity that facilitates revision through reuse. In this paper we are concerned with requirements that are extensional in the sense that they pertain directly to observable behavior. For a simple example, consider a program acting on variables hi, lo where the initial value of hi is meant to be a secret, on which the final value of lo must not depend. Consider this simple notion of program behavior: a state assigns values to variables, and a trace is a pair: the initial and final states. The requirement cannot be specified as a trace property, but it can be specified as follows: for any two traces (σ, σ') and (τ, τ'), if the initial states σ and τ have the same value for lo then so do the final states. In symbols: $\sigma(lo) = \tau(lo) \Rightarrow \sigma'(lo) = \tau'(lo)$.

Some requirements involve more than two traces, e.g., "the average response time is under a millisecond" can be made precise by averaging the response time of each trace, over all traces, perhaps weighted by a distribution that represents likelihood of different requests. For a non-quantitative example, consider the requirement that a process in a distributed system should know which process is the leader: something is known in a given execution if it is true in all possible traces that are consistent with what the process can observe of the given execution (such as a subset of the network messages). In the security literature, some information flow properties are defined by closure conditions on the program's traces, such as: for any two traces, there exists a trace with the high (confidential) events of the first and the low (public) events of the second.

This paper explores the notion that just as a property of books is a set of books, not necessarily defined simply in terms of their sentences, so too a property of programs is a set of programs, not merely a set of traces. The goal is to investigate how the algebra of programming can be adapted for reasoning about non-trace properties. To this end, we focus on the most rudimentary notion of trace, i.e., pre/post pairs, and rudimentary program constructs. We conjecture that the phenomena and ideas are relevant to a range of models, perhaps even the rich notions of trace abstracted by variations on concurrent Kleene algebra [20].

It is unfortunate that the importance of trace properties in programming has led to well established use of the term "property" for trace property, and recent escalation in terminology to "hyperproperty" to designate the general notion of program property—sets of programs rather than sets of traces [9,10]. Some distinction is needed, so for clarity and succinctness we follow the crowd.

---

[1] This paper was written with the UTP [19] community in mind, but our use of the term "design" is informal and does not refer to the technical notion in UTP.

The technical contribution of this paper can now be described as follows: we give a lifting of the fixpoint semantics of loops to the "hyper level", and show anomalies that occur with other liftings. This enables reasoning at the hyper level with usual fixpoint laws for loops, while retaining consistency with standard relational semantics. Rather than working directly with sets of trace sets, our lifting uses a simpler model, sets of state sets; this serves to illustrate the issues and make connections with other models that may be familiar. The conceptual contribution of the paper is to call attention to the challenge of unifying theories of programming that encompass requirements beyond trace properties.

Outline. Section 2 describes a relational semantics of imperative programs and defines an example program property that is not a trace property. Relational semantics is connected, in Sect. 3, with semantics mapping sets to sets, like forward predicate transformers. Section 4 considers semantics mapping sets of sets to the same, this being the level at which hyperproperties are formulated. Anomalies with obvious definitions motivate a more nuanced semantics of loops. The main technical result of the paper is Theorem 1 in this section, connecting the semantics of Sect. 4 with that of Sect. 3. Section 5 connects the preceding results with the intrinsic notion of satisfaction for hyperproperties, and sketches challenges in realizing the dream of reasoning about hyperproperties using only refinement chains. The semantics and theorem are new, but similar to results in prior work discussed in Sect. 6. Section 7 concludes.

2   Programs and Specifications as Binary Relations

Preliminaries. We review some standard notions, to fix notation and set the stage. Throughout the paper we assume Σ is a nonempty set, which stands for the set of program states, or data values, on which programs act. For any sets A, B, let $A \multimap B$ denote the binary relations from A to B; that is, $A \multimap B$ is $\wp(A \times B)$ where $\wp$ means powerset. Unless otherwise mentioned, we consider powersets, including $\Sigma \multimap \Sigma$, to be ordered by inclusion ($\subseteq$).

We write $A \to B$ for the set of functions from A to B. For composition of relations, and in particular composition of functions, we use infix symbol ; in the forward direction. Thus for relations R, S and elements x, y we have $x(R \,;\, S)y$ iff $\exists z \bullet xRz \wedge zSy$. For a function $f : A \to B$ and element $x \in A$ we write application as fx and let it associate to the left. Composition with $g : B \to C$ is written $f \,;\, g$, as functions are treated as special relations, so $(f \,;\, g)x = g(fx)$. The symbol ; binds tighter than $\cup$ and other operators.

For a relation $R : A \multimap B$, the direct image $\langle R \rangle$ is a total function $\wp A \to \wp B$ defined by $y \in \langle R \rangle p$ iff $\exists x \in p \bullet xRy$. It faithfully reflects ordering of relations:

$$R \subseteq S \quad \text{iff} \quad \langle R \rangle \sqsubseteq \langle S \rangle$$

where $\sqsubseteq$ means pointwise order (i.e., $\varphi \sqsubseteq \psi$ iff $\forall p \in \wp A \bullet \varphi p \subseteq \psi p$). We write $\sqcup$ for pointwise union, defined by $(\varphi \sqcup \psi)p = \varphi p \cup \psi p$. The $\sqsubseteq$-least element is the function $\lambda p \bullet \emptyset$, abbreviated as $\bot$. A relation can be recovered from its direct image:

$$R = sglt \, ; \, \langle R \rangle \, ; \, \ni \tag{1}$$

where $sglt : A \to \wp A$ maps element a to singleton set $\{a\}$ and $\ni \, : \, \wp A \multimap A$ is the converse of the membership relation. Note that $\bot$ is the direct image of the empty relation. Direct image is functorial and distributes over union:

$$\langle id_\Sigma \rangle = id_{\wp \Sigma} \qquad \langle R \, ; \, S \rangle = \langle R \rangle \, ; \, \langle S \rangle \qquad \langle R \cup S \rangle = \langle R \rangle \sqcup \langle S \rangle$$

We write id for identity function on the set indicated. In fact $\langle - \rangle$ distributes over arbitrary union, i.e., sends any union of relations to the pointwise join of their images. Also, $\langle R \rangle$ is universally disjunctive, and (1) forms a bijection between universally disjunctive functions $\wp A \to \wp B$ and relations $A \multimap B$.

In this paper we use the term **transformer** for monotonic functions of type $\wp A \to \wp B$. For $\varphi : \wp A \to \wp B$ to be monotonic is equivalent to $(\supseteq \, ; \, \varphi) \subseteq (\varphi \, ; \, \supseteq)$.

We write lfp for the least-fixpoint operator. For monotonic functions $f : A \to A$ and $g : B \to B$ where A, B are sufficiently complete posets that lfp f and lfp g exist, the **fixpoint fusion** rule says that for strict and continuous $h : A \to B$,

$$f \, ; \, h = h \, ; \, g \Rightarrow h(\text{lfp } f) = \text{lfp } g \tag{2}$$

Inequational forms, such as $f \, ; \, h \leq h \, ; \, g \Rightarrow h(\text{lfp } f) \leq \text{lfp } g$, are also important.[2]

Relational Semantics. The relational model suffices for reasoning about terminating executions. If we write $x + 2 \leq x'$ to specify a program that increases x by at least two, we can write this simple refinement chain:

$$x + 2 \leq x' \quad \supseteq \quad x := x + 3 \oplus x := x + 5 \quad \supseteq \quad x := x + 3$$

to express that the nondeterministic choice ($\oplus$) between adding 3 or adding 5 refines the specification and is refined in turn by the first alternative. Relations model a good range of operations including relational converse and intersection which are not implementable in general but are useful for expressing specifications. Their algebraic laws facilitate reasoning. For example, choice is modeled as union, so the second step is from a law of set theory: $R \cup S \supseteq R$.

Equations and inequations may serve as specifications. For example, to express that relation R is deterministic we can write $R^\cup \, ; \, R \subseteq id$, where $R^\cup$ is the converse of R. Note that this uses two occurrences of R. Returning to the example in the introduction, suppose R relates states with variables hi, lo. To formulate the **noninterference** property that the final value of lo is independent of the initial value of hi, it is convenient to define a relation on states that says they have the same value for lo: define $\overset{\circ}{\sim}$ by $\sigma \overset{\circ}{\sim} \tau$ iff $\sigma(lo) = \tau(lo)$. The property is

$$\forall \sigma, \sigma', \tau, \tau' \bullet \sigma R \sigma' \wedge \tau R \tau' \wedge \sigma \overset{\circ}{\sim} \tau \Rightarrow \sigma' \overset{\circ}{\sim} \tau'$$

---

[2] Fusion rules, also called fixpoint transfer, can be found in many sources, e.g., [1,4]. We need the form in Theorem 3 of [12], for Kleene approximation of fixpoints.

This is a form of determinacy. A weaker notion allows multiple outcomes for lo but the set of possibilities should be independent from the initial value of hi.

$$\forall \sigma, \sigma', \tau \bullet \sigma R \sigma' \wedge \sigma \overset{\sim}{\approx} \tau \Rightarrow \exists \tau' \bullet \tau R \tau' \wedge \sigma' \overset{\sim}{\approx} \tau'$$

This is known as **possibilistic noninterference**. It can be expressed without quantifiers, by the usual simulation inequality:

$$\overset{\sim}{\approx} ; R \subseteq R ; \overset{\sim}{\approx} \tag{3}$$

Another equivalent form is $\overset{\sim}{\approx} ; R ; \overset{\sim}{\approx} = R ; \overset{\sim}{\approx}$, which again uses two occurrences of R. The algebraic formulations are attractive, but recall the beautiful idea of correctness proof as a chain of refinements

$$spec \supseteq design \supseteq \ldots \supseteq prog$$

This requires the specification to itself be a term in the algebra, rather than an (in)equation between terms.

Before proceeding to investigate this issue, we recall the well known fact that possibilistic noninterference is not closed under refinement of trace sets [21]. Consider hi, lo ranging over bits, so we can write pairs compactly, and consider the set of traces $\{(00, 00), \underline{(00, 01)}, (01, 00), \underline{(01, 01)}, (10, 10), (10, 11), \underline{(11, 10)}, (11, 11)\}$ It satisfies possibilistic noninterference, but if we remove the underlined pairs the result does not; in fact the result copies hi to lo.

In the rest of this paper, we focus on deterministic noninterference, NI for short. It has been advocated as a good notion for security [35] and it serves our purposes as an example.

A Signature and Its Relational Model. To investigate how NI and other non-trace properties may be expressed and used in refinement chains, it is convenient to focus on a specific signature, the simple imperative language over given atoms (ranged over by atm) and boolean expressions (ranged over by b).

$$c ::= atm \mid \mathsf{skip} \mid c; c \mid c \oplus c \mid \mathsf{if}\ b\ \mathsf{then}\ c\ \mathsf{else}\ c \mid \mathsf{while}\ b\ \mathsf{do}\ c \tag{4}$$

For expository purposes we refrain from decomposing the conditional and iteration constructs in terms of choice ($\oplus$) and assertions. That decomposition would be preferred in a more thorough investigation of algebraic laws, and it is evident in the semantic definitions to follow.

Assume that for each atm is given a relation $[\![atm]\!] : \Sigma \multimap \Sigma$, and for each boolean expression b is given a coreflexive relation $[\![b]\!] : \Sigma \multimap \Sigma$. That is, $[\![b]\!]$ is a subset of the identity relation id_Σ on Σ. For non-atom commands c the relational semantics $[\![c]\!]$ is defined in Fig. 1. The fixpoint for loops[3] is in $\Sigma \multimap \Sigma$, ordered by $\subseteq$ with least element $\emptyset$.

---

[3] It is well known that loops are expressible in terms of recursion: while b do c can be expressed as $\mu X.(b; c; X \cup \neg b)$ and this is the form we use in semantics. A well known law is $\mu X.(b; c; X \cup \neg b) = \mu X.(b; c; X \cup \mathsf{skip}); \neg b$ which factors out the termination condition.

$$\begin{aligned}
[\![\, \mathsf{skip}\,]\!] &= id_\Sigma \\
[\![\, c\,;d\,]\!] &= [\![\, c\,]\!]\,;[\![\, d\,]\!] \\
[\![\, c \oplus d\,]\!] &= [\![\, c\,]\!] \cup [\![\, d\,]\!] \\
[\![\, \mathsf{if}\ b\ \mathsf{then}\ c\ \mathsf{else}\ d\,]\!] &= [\![\, b\,]\!]\,;[\![\, c\,]\!]\ \cup\ [\![\, \neg b\,]\!]\,;[\![\, d\,]\!] \\
[\![\, \mathsf{while}\ b\ \mathsf{do}\ c\,]\!] &= \mathsf{lfp}\,\mathsf{F}
\end{aligned}$$

where $\mathsf{F} : (\Sigma \multimap \Sigma) \to (\Sigma \multimap \Sigma)$ is defined

$$\mathsf{F}R = [\![\, b\,]\!]\,;[\![\, c\,]\!]\,;R \cup [\![\, \neg b\,]\!]$$

Fig. 1. Relational semantics $[\![\, c\,]\!] \in \Sigma \multimap \Sigma$, with $[\![\, atm\,]\!]$ assumed to be given.

$$\begin{aligned}
\{\!|\, atm\,|\!\} &= \langle [\![\, atm\,]\!] \rangle \\
\{\!|\, \mathsf{skip}\,|\!\} &= id_{\wp\Sigma} \\
\{\!|\, c\,;d\,|\!\} &= \{\!|\, c\,|\!\}\,;\{\!|\, d\,|\!\} \\
\{\!|\, c \oplus d\,|\!\} &= \{\!|\, c\,|\!\} \sqcup \{\!|\, d\,|\!\} \\
\{\!|\, \mathsf{if}\ b\ \mathsf{then}\ c\ \mathsf{else}\ d\,|\!\} &= \{\!|\, b\,|\!\}\,;\{\!|\, c\,|\!\}\ \sqcup\ \{\!|\, \neg b\,|\!\}\,;\{\!|\, d\,|\!\} \\
\{\!|\, \mathsf{while}\ b\ \mathsf{do}\ c\,|\!\} &= \mathsf{lfp}\,\mathsf{G}
\end{aligned}$$

where $\mathsf{G} : (\wp\Sigma \to \wp\Sigma) \to (\wp\Sigma \to \wp\Sigma)$ is defined

$$\mathsf{G}\varphi = \{\!|\, b\,|\!\}\,;\{\!|\, c\,|\!\}\,;\varphi \sqcup \{\!|\, \neg b\,|\!\}$$

Fig. 2. Transformer semantics $\{\!|\, c\,|\!\} \in \wp\Sigma \to \wp\Sigma$.

The language goes beyond ordinary programs, in the sense that atoms are allowed to be unboundedly nondeterministic. They are also allowed to be partial; coreflexive atoms serve as assume and assert statements. Other ingredients are needed for a full calculus of specifications, but here our aim is to sketch ideas that merit elaboration in a more comprehensive theory.

3   Programs as Forward Predicate Transformers

Here is yet another way to specify NI for a relation R:

$$\forall p \in \wp\Sigma \bullet \mathsf{AgrI}\,(p) \Rightarrow \mathsf{AgrI}\,(\langle R\rangle p)$$

where AgrI says that all elements of p agree on lo:

$$\mathsf{AgrI}\,(p) \quad \text{iff} \quad \forall \sigma, \tau \bullet \sigma \in p \wedge \tau \in p \Rightarrow \sigma \overset{\circ}{\sim} \tau$$

As with the preceding (in)equational formulations, like (3), this is not directly applicable as the specification in a refinement chain, but it does hint that escalating to sets of states may be helpful. Note that R occurs just once in the condition.

Weakest-precondition predicate transformers are a good model for programming algebra: Monotonic functions $\wp\Sigma \to \wp\Sigma$ can model total correctness specifications with both angelic and demonic nondeterminacy. In this paper we use transformers to model programs in the forward direction.

For boolean expression b we define $\{\!\mid b \mid\!\} = \langle\![\, b\,]\!\rangle$ so that $\{\!\mid b \mid\!\}$ is a filter: x is in $\{\!\mid b \mid\!\}p$ iff $x \in p$ and b is true of x. The transformer semantics is in Fig. 2. For loops, the fixpoint is for the aforementioned $\bot$ and $\sqsubseteq$.

Linking Transformer With Relational. The transformer model may support a richer range of operators than the relational one, but for several reasons it is important to establish their mutual consistency on a common set of operators [18, 19]. A relation can be recovered from its direct image, see (1), so the following is a strong link.

Proposition 1. *For all c in the signature,* $\langle\![\,[\![c]\!] \,]\!\rangle = \{\!\mid c \mid\!\}$.

Proof. By induction on c.

- skip: $\langle\![\, [\![\text{skip}]\!] \,]\!\rangle = \langle id_\Sigma \rangle = id_{\wp\Sigma \to \wp\Sigma} = \{\!\mid \text{skip} \mid\!\}$ by definitions and $\langle - \rangle$ law.
- *atm*: $\langle\![\, [\![atm]\!] \,]\!\rangle = \{\!\mid atm \mid\!\}$ by definition.
- $c; d$: $\langle\![\, [\![c; d]\!] \,]\!\rangle = \langle\![\, [\![c]\!]; [\![d]\!] \,]\!\rangle = \langle\![\, [\![c]\!] \,]\!\rangle; \langle\![\, [\![d]\!] \,]\!\rangle = \{\!\mid c \mid\!\}; \{\!\mid d \mid\!\} = \{\!\mid c; d \mid\!\}$ by definitions, $\langle - \rangle$ laws, and induction hypothesis.
- $c \oplus d$: $\langle\![\, [\![c \oplus d]\!] \,]\!\rangle = \langle\![\, [\![c]\!] \cup [\![d]\!] \,]\!\rangle = \langle\![\, [\![c]\!] \,]\!\rangle \sqcup \langle\![\, [\![d]\!] \,]\!\rangle = \{\!\mid c \mid\!\} \sqcup \{\!\mid d \mid\!\} = \{\!\mid c \oplus d \mid\!\}$ by definitions, $\langle - \rangle$ laws, and induction hypothesis.
- if b then c else d: $\langle\![\, [\![\text{if } b \text{ then } c \text{ else } d]\!] \,]\!\rangle = \langle\![\, [\![b]\!] ; [\![c]\!] \cup [\![\neg b]\!] ; [\![d]\!] \,]\!\rangle = \langle\![\, [\![b]\!] \,]\!\rangle ; \langle\![\, [\![c]\!] \,]\!\rangle \sqcup \langle\![\, [\![\neg b]\!] \,]\!\rangle ; \langle\![\, [\![d]\!] \,]\!\rangle = \{\!\mid b \mid\!\}; \langle\![\, [\![c]\!] \,]\!\rangle \sqcup \{\!\mid \neg b \mid\!\}; \langle\![\, [\![d]\!] \,]\!\rangle = \{\!\mid b \mid\!\}; \{\!\mid c \mid\!\} \sqcup \{\!\mid \neg b \mid\!\}; \{\!\mid d \mid\!\} = \{\!\mid \text{if } b \text{ then } c \text{ else } d \mid\!\}$ by definitions, $\langle - \rangle$ laws, and induction hypothesis.
- while b do c: To prove $\langle\![\, [\![\text{while } b \text{ do } c]\!] \,]\!\rangle = \{\!\mid \text{while } b \text{ do } c \mid\!\}$, unfold the definitions to $\langle \text{lfp}\, F \rangle = \text{lfp}\, G$, where F, G are defined in Figs. 1 and 2. This follows by fixpoint fusion, taking h in (2) to be $\langle - \rangle$ so the antecedent to be proved is $\forall R \bullet \langle F R \rangle = G \langle R \rangle$. Observe for any R:

$$
\begin{aligned}
&\langle F R \rangle \\
={}& \langle\![\, [\![b]\!] ; [\![c]\!] ; R \cup [\![\neg b]\!] \,]\!\rangle && \text{def } F \\
={}& \langle\![\, [\![b]\!] \,]\!\rangle ; \langle\![\, [\![c]\!] \,]\!\rangle ; \langle R \rangle \sqcup \langle\![\, [\![\neg b]\!] \,]\!\rangle && \langle - \rangle \text{ distributes over ; and } \cup \\
={}& \{\!\mid b \mid\!\}; \langle\![\, [\![c]\!] \,]\!\rangle ; \langle R \rangle \sqcup \{\!\mid \neg b \mid\!\} && \text{def } \{\!\mid b \mid\!\} \\
={}& \{\!\mid b \mid\!\}; \{\!\mid c \mid\!\}; \langle R \rangle \sqcup \{\!\mid \neg b \mid\!\} && \text{induction hypothesis} \\
={}& G \langle R \rangle && \text{def } G
\end{aligned}
$$

$\square$

Subsets of $\wp\Sigma \to \wp\Sigma$, such as transformers satisfying Dijkstra's healthiness conditions, validate stronger laws than the full set of (monotonic) transformers. Healthiness conditions can be expressed by inequations, such as the determinacy inequation $R^\cup ; R \subseteq id$, and used as antecedents in algebraic laws. Care must be taken with joins: not all subsets are closed under pointwise union. Pointwise union does provide joins in the set of all transformers and also in the set of all universally disjunctive transformers.

$$
\begin{aligned}
(\!| \, atm \, |\!) &= \langle \{\!| \, atm \, |\!\} \rangle \\
(\!| \, \text{skip} \, |\!) &= id \\
(\!| \, c \, ; d \, |\!) &= (\!| \, c \, |\!) \, ; (\!| \, d \, |\!) \\
(\!| \, c \oplus d \, |\!) &= (\!| \, c \, |\!) \, \otimes \, (\!| \, d \, |\!) \\
(\!| \, \text{if } b \text{ then } c \text{ else } d \, |\!) &= (\!| \, c \, |\!) \lhd b \rhd (\!| \, d \, |\!) \\
(\!| \, \text{while } b \text{ do } c \, |\!) &= \text{lfp H}
\end{aligned}
$$

where $\mathsf{H} : (\breve{\wp}(\wp\Sigma) \to \breve{\wp}(\wp\Sigma)) \to (\breve{\wp}(\wp\Sigma) \to \breve{\wp}(\wp\Sigma))$ is defined
$\mathsf{H}\Phi = (\!| \, c \, |\!) \, ; \Phi \lhd b \rhd (\!| \, \text{skip} \, |\!)$

Fig. 3. H-transformer semantics $(\!| \, c \, |\!) \in \breve{\wp}(\wp\Sigma) \to \breve{\wp}(\wp\Sigma)$.

In addition to transformers as weakest preconditions [4], another similar model is multirelations which are attractive in maintaining a pre-to-post direction [25]. These are all limited to trace properties, though, so we proceed in a different direction.

4   Programs as h-Transformers

Given $R : A \multimap B$, the image $\langle R \rangle$ is a function and functions are relations, so the direct image can be taken: $\langle\langle R \rangle\rangle : \wp^2 A \to \wp^2 B$ where $\wp^2 A$ abbreviates $\wp(\wp A)$. In this paper, monotonic functions of this type are called **h-transformers**, in a nod to hyper terminology.

The underlying relation can be recovered by two applications of (1):

$$
R = sglt \, ; sglt \, ; \langle\langle R \rangle\rangle \, ; \ni \, ; \ni
$$

More to the point, a quantifier-free formulation of NI is now in reach. Recall that we have $R \in \mathsf{NI}$ iff $\forall p \in \wp\Sigma \bullet \mathsf{Agrl}\,(p) \Rightarrow \mathsf{Agrl}\,(\langle R \rangle p)$. This is equivalent to

$$
\langle\langle R \rangle\rangle \mathbb{A} \subseteq \mathbb{A} \tag{5}
$$

where the set of sets $\mathbb{A}$ is defined by $\mathbb{A} = \{p \mid \mathsf{Agrl}\,(p)\}$. This is one motivation to investigate $\wp^2\Sigma \to \wp^2\Sigma$ as a model, rather than $\wp(\Sigma \multimap \Sigma)$ which is the obvious way to embody the idea that a program is a trace set and a property is a set of programs.

In the following we continue to write $\sqcup$ and $\sqsubseteq$ for the pointwise join and pointwise order on $\wp^2\Sigma \to \wp^2\Sigma$. Please note the order is defined in terms of set inclusion at the outer layer of sets and is independent of the order on $\wp\Sigma$. Define $\bot = \langle \bot \rangle$ and note that $\bot\emptyset = \emptyset$ and $\bot\mathbb{Q} = \{\emptyset\}$ for $\mathbb{Q} \neq \emptyset$.

Surprises. For semantics using h-transformers, some obvious guesses work fine but others do not. The semantics in Fig. 3 uses operators $\otimes$, $\lhd b \rhd$ and $\breve{\wp}$ which will be explained in due course. For boolean expressions we simply lift by direct image, defining $(\!| \, b \, |\!) = \langle \{\!| \, b \, |\!\} \rangle$. The same for command atoms, so the semantics of atm is derived from the given $[\![\, atm \,]\!]$.

The analog of Proposition 1 is that for all c in the signature, $\langle\!\{\!|c|\!\}\rangle = (\!|c|\!)$, allowing laws valid in relational semantics to be lifted to h-transformers. Considering some cases suggests that this could be proved by induction on c:

- skip: $\langle\!\{\!|\,\mathsf{skip}\,|\!\}\rangle = \langle id_{\wp\Sigma}\rangle = (\!|\,\mathsf{skip}\,|\!)$ by definitions and using that $\langle-\rangle$ preserves identity.
- atm: $\langle\!\{\!|\,atm\,|\!\}\rangle = (\!|\,atm\,|\!)$ by definition.
- $c;d$: $\langle\!\{\!|\,c;d\,|\!\}\rangle = \langle\!\{\!|c|\!\}\,;\{\!|d|\!\}\rangle = \langle\!\{\!|c|\!\}\rangle\,;\langle\!\{\!|d|\!\}\rangle = (\!|c|\!)\,;(\!|d|\!) = (\!|c;d|\!)$ by definitions, distribution of $\langle-\rangle$ over ;, and putative induction hypothesis.

These calculations suggest we may succeed with this obvious guess:

$$(\!|\,\mathsf{if}\ b\ \mathsf{then}\ c\ \mathsf{else}\ d\,|\!) = (\!|b|\!)\,;(\!|c|\!) \sqcup (\!|\neg b|\!)\,;(\!|d|\!) \tag{6}$$

The induction hypothesis would give $(\!|\,\mathsf{if}\ b\ \mathsf{then}\ c\ \mathsf{else}\ d\,|\!) = \langle\!\{\!|b|\!\}\,;\{\!|c|\!\}\rangle\sqcup\langle\!\{\!|\neg b|\!\}\,;\{\!|d|\!\}\rangle$. On the other hand, $\langle\!\{\!|\,\mathsf{if}\ b\ \mathsf{then}\ c\ \mathsf{else}\ d\,|\!\}\rangle = \langle\!\{\!|b|\!\}\,;\{\!|c|\!\}\sqcup\{\!|\neg b|\!\}\,;\{\!|d|\!\}\rangle$. Unfortunately these are quite different because the joins are at different levels. In general, for φ and ψ of type $\wp\Sigma \to \wp\Sigma$ and $\mathbb{Q} \in \wp^2\Sigma$ we have $\langle\varphi\sqcup\psi\rangle\mathbb{Q} = \{\varphi p\cup\psi p \mid p \in \mathbb{Q}\}$ whereas $(\langle\varphi\rangle\sqcup\langle\psi\rangle)\mathbb{Q} = \{\varphi p \mid p \in \mathbb{Q}\}\cup\{\psi p \mid p \in \mathbb{Q}\}$. Indeed, the same discrepancy would arise if we define $(\!|c\oplus d|\!) = (\!|c|\!)\sqcup(\!|d|\!)$.

At this point one may investigate notions of "inner join", but for expository purposes we proceed to consider a putative definition for loops. Following the pattern for relational and transformer semantics, an obvious guess is

$$(\!|\,\mathsf{while}\ b\ \mathsf{do}\ c\,|\!) = \mathsf{lfp}\,\mathsf{K}\ \text{where}\ \mathsf{K}\Phi = (\!|b|\!)\,;(\!|c|\!)\,;\Phi\sqcup(\!|\neg b|\!) \tag{7}$$

Consider this program: while $x < 4$ do $x := x + 1$. We can safely assume $(\!|\,x < 4\,|\!)$ is $\langle\!\{\!|\,x < 4\,|\!\}\rangle$ and $(\!|\,x := x+1\,|\!)$ is $\langle\!\{\!|\,x := x+1\,|\!\}\rangle$. As there is a single variable, we can represent a state by its value, for example $\{2,5\}$ is a set of two states. Let us work out $(\!|\,\mathsf{while}\ x < 4\ \mathsf{do}\ x := x + 1\,|\!)\{\{2,5\}\}$. Now $(\!|\,\mathsf{while}\ x < 4\ \mathsf{do}\ x := x + 1\,|\!)$ is the limit of the chain $\mathsf{K}^i\bot$ where K^i means i applications of K. Note that for any Φ and $i > 0$,

$$\mathsf{K}^i\Phi = ((\!|x < 4|\!)\,;(\!|x := x+1|\!))^i\,;\Phi\ \sqcup$$
$$(\sqcup j :: 0 \le j < i \bullet ((\!|x < 4|\!)\,;(\!|x := x+1|\!))^j\,;(\!|\neg x < 4|\!))$$

Writing $\mathbb{Q}_i$ for $\mathsf{K}^i\bot\{\{2,5\}\}$ one can derive

$$\mathbb{Q}_0 = \{\emptyset\}$$
$$\mathbb{Q}_1 = \{\emptyset\} \cup \{\{5\}\} = \{\emptyset, \{5\}\}$$
$$\mathbb{Q}_2 = \{\emptyset\} \cup \{\emptyset\} \cup \{\{5\}\} = \{\emptyset, \{5\}\}$$
$$\mathbb{Q}_3 = \{\emptyset\} \cup \{\emptyset\} \cup \{\{4\}\} \cup \{\{5\}\} = \{\emptyset, \{4\}, \{5\}\}$$

at which point the sequence remains fixed. As in the case of conditional (6), the result is not consistent with the underlying semantics:

$$\{\!|\,\mathsf{while}\ x < 4\ \mathsf{do}\ x := x + 1\,|\!\}\{2,5\} = \{4,5\}$$

The result should be $\{\{4,5\}\}$ if we are to have the analog of Proposition 1.

A plausible inner join is $\otimes$ defined by $(\Phi \otimes \Psi)\mathbb{Q} = \{r \cup s \mid \exists q \in \mathbb{Q} \bullet r \in \Phi\{q\} \wedge s \in \Psi\{q\}\}$. This can be used to define a semantics of $\oplus$ as well as semantics of conditional and loop; the resulting constructs are $\sqsubseteq$-monotonic and enjoy other nice properties.

Indeed, using $\otimes$ in place of $\sqcup$ in (7), we get $\mathsf{K}^3 \bot\{\{2,5\}\} = \{\{4,5\}\}$, which is exactly the lift of the transformer semantics. There is one serious problem: K fails to be increasing. In particular, $\bot \not\sqsubseteq \mathsf{K}\bot$; for example $\bot\{\{2,5\}\} = \{\emptyset\}$ but $\mathsf{H}\bot\{\{2,5\}\} = \{\{5\}\}$. While this semantics merits further study, we leave it aside because we aim to use fixpoint fusion results that rely on Kleene approximation: This requires $\bot \sqsubseteq \mathsf{K}\bot$ in order to have an ascending chain, and the use of $\bot$ so that $\langle\ \rangle$ is strict.

A Viable Solution. Replacing singleton by powerset in the definition of $\otimes$, for any h-transformers $\Phi, \Psi : \wp^2\Sigma \to \wp^2\Sigma$ we define the inner join $\oslash$ by

$$(\Phi \oslash \Psi)\mathbb{Q} = \{r \cup s \mid \exists p \in \mathbb{Q} \bullet r \in \Phi(\wp\, p) \wedge s \in \Psi(\wp\, p)\}$$

For semantics of conditionals, it is convenient to define, for boolean expression b, this operator on h-transformers: $\Phi \triangleleft b \triangleright \Psi = (\!|b|\!)\,;\Phi \oslash (\!|\neg b|\!)\,;\Psi$. It satisfies

$$(\Phi \triangleleft b \triangleright \Psi)\mathbb{Q} = \{r \cup s \mid \exists p \in \mathbb{Q} \bullet r \in \Phi(\wp(\{\!|b|\!\}p)) \wedge s \in \Psi(\wp(\{\!|\neg b|\!\}p))\} \quad (8)$$

because $(\!|b|\!)(\wp\, p) = \langle\{\!|b|\!\}\rangle(\wp\, p) = \wp(\{\!|b|\!\}p))$. These operators are used in Fig. 3 for semantics of conditional and loop.

It is straightforward to prove $\oslash$ is monotonic: $\Phi \sqsubseteq \Phi'$ and $\Psi \sqsubseteq \Psi'$ imply $\Phi \oslash \Psi \sqsubseteq \Phi' \oslash \Psi'$. It is also straightforward to prove

$$\langle \varphi \sqcup \psi \rangle \sqsubseteq \langle \varphi \rangle \oslash \langle \psi \rangle \quad (9)$$

but in general equality does not hold, so we focus on $\triangleleft - \triangleright$.

Lemma 1. *For any b, $\triangleleft\, b\, \triangleright$ is monotonic: $\Phi \sqsubseteq \Phi'$ and $\Psi \sqsubseteq \Psi'$ imply $\Phi \triangleleft\, b\, \triangleright \Psi \sqsubseteq \Phi' \triangleleft\, b\, \triangleright \Psi'$.*

Proof. Keep in mind this is $\sqsubseteq$ at the outer level: $\Phi \sqsubseteq \Phi'$ means $\forall \mathbb{Q} \bullet \Phi\mathbb{Q} \subseteq \Phi'\mathbb{Q}$ (more sets, not bigger sets, if you will). This follows by monotonicity of $\oslash$, or using characterization (8) we have

$$r \cup s \in (\Phi \triangleleft\, b\, \triangleright \Psi)\mathbb{Q} \quad \text{iff} \quad \exists q \in \mathbb{Q} \bullet r \in \Phi(\wp(\{\!|b|\!\}q)) \wedge s \in \Psi(\wp(\{\!|\neg b|\!\}q))$$

which implies $\exists q \in \mathbb{Q} \bullet r \in \Phi'(\wp(\{\!|b|\!\}q)) \wedge s \in \Psi'(\wp(\{\!|\neg b|\!\}q))$ by $\Phi \sqsubseteq \Phi'$ and $\Psi \sqsubseteq \Psi'$. $\quad\square$

With $(\!|\text{if } b \text{ then } c \text{ else } d|\!)$ defined as in Fig. 3 we have the following refinement.

Lemma 2. $\langle\{\!|\text{if } b \text{ then } c \text{ else } d|\!\}\rangle \sqsubseteq (\!|\text{if } b \text{ then } c \text{ else } d|\!)$ *provided that* $\langle\{\!|c|\!\}\rangle \sqsubseteq (\!|c|\!)$ *and* $\langle\{\!|d|\!\}\rangle \sqsubseteq (\!|d|\!)$.

Proof.

$$\langle\{\!\!\{\,\text{if } b \text{ then } c \text{ else } d\,\}\!\!\}\rangle$$
$$=\qquad\qquad\qquad\qquad\qquad\qquad\text{semantics}$$
$$\langle\{\!\!\{\,b\,\}\!\!\}\,;\,\{\!\!\{\,c\,\}\!\!\}\sqcup\{\!\!\{\,\neg b\,\}\!\!\}\,;\,\{\!\!\{\,d\,\}\!\!\}\rangle$$
$$\sqsubseteq\qquad\qquad\qquad\qquad\qquad\quad\text{by (9)}$$
$$\langle\{\!\!\{\,b\,\}\!\!\}\,;\,\{\!\!\{\,c\,\}\!\!\}\rangle\otimes\langle\{\!\!\{\,\neg b\,\}\!\!\}\,;\,\{\!\!\{\,d\,\}\!\!\}\rangle$$
$$=\qquad\qquad\qquad\qquad\qquad\quad\text{distribute }\langle-\rangle\text{ over };,\text{ semantics}$$
$$(\!|\,b\,|\!)\,;\langle\{\!\!\{\,c\,\}\!\!\}\rangle\otimes(\!|\,\neg b\,|\!)\,;\langle\{\!\!\{\,d\,\}\!\!\}\rangle$$
$$\sqsubseteq\qquad\qquad\qquad\qquad\qquad\quad\text{assumption, monotonicity}$$
$$(\!|\,b\,|\!)\,;(\!|\,c\,|\!)\otimes(\!|\,\neg b\,|\!)\,;(\!|\,d\,|\!)$$
$$=\qquad\qquad\qquad\qquad\qquad\quad\text{semantics, def of }\lhd\,b\,\rhd\text{ from }\otimes$$
$$(\!|\,\text{if } b \text{ then } c \text{ else } d\,|\!)$$

$$\square$$

This result suggests that we might be able to prove $\langle\{\!\!\{\,c\,\}\!\!\}\rangle\sqsubseteq(\!|\,c\,|\!)$ for all c, but that would be a weak link between the transformer and h-transformer semantics. A stronger link can be forged as follows.

We say $\mathbb{Q}\in\wp^2\Sigma$ is **subset closed** iff $\mathbb{Q}=\mathsf{ssc}\,\mathbb{Q}$ where the subset closure operator ssc is defined by $p\in\mathsf{ssc}\,\mathbb{Q}$ iff $\exists q\in\mathbb{Q}\bullet p\subseteq q$. For example, the set $\mathbb{A}$ used in (5) is subset closed. Observe that $\mathsf{ssc}=\langle\supseteq\rangle$.

Lemma 3. *For transformers $\varphi,\psi:\wp\Sigma\to\wp\Sigma$ and condition b, if $\mathbb{Q}=\mathsf{ssc}\,\mathbb{Q}$ then $\langle\{\!\!\{\,b\,\}\!\!\}\,;\,\varphi\sqcup\{\!\!\{\,\neg b\,\}\!\!\}\,;\,\psi\rangle\mathbb{Q}=(\langle\varphi\rangle\lhd\,b\,\rhd\langle\psi\rangle)\mathbb{Q}$.*

Proof. For the LHS, by definitions:

$$\langle\{\!\!\{\,b\,\}\!\!\}\,;\,\varphi\sqcup\{\!\!\{\,\neg b\,\}\!\!\}\,;\,\psi\rangle\mathbb{Q}$$
$$=$$
$$\{r\cup s\mid\exists q\in\mathbb{Q}\bullet r=\varphi(\{\!\!\{\,b\,\}\!\!\}q)\wedge s=\psi(\{\!\!\{\,\neg b\,\}\!\!\}q)\}\qquad(*)$$

For the RHS, again by definitions:

$$(\langle\varphi\rangle\lhd\,b\,\rhd\langle\psi\rangle)\mathbb{Q}$$
$$=$$
$$\{r\cup s\mid\exists q\in\mathbb{Q}\bullet r\in\langle\varphi\rangle(\wp(\{\!\!\{\,b\,\}\!\!\}q))\wedge s\in\langle\psi\rangle(\wp(\{\!\!\{\,\neg b\,\}\!\!\}q))\}$$
$$=$$
$$\{r\cup s\mid\exists q\in\mathbb{Q}\bullet\exists t,u\bullet t\subseteq\{\!\!\{\,b\,\}\!\!\}q\wedge u\subseteq\{\!\!\{\,\neg b\,\}\!\!\}q\wedge r=\varphi t\wedge s=\psi u\}\qquad(\dagger)$$

Now $(*)\subseteq(\dagger)$ by instantiating $t:=\{\!\!\{\,b\,\}\!\!\}q$ and $u:=\{\!\!\{\,\neg b\,\}\!\!\}q$, so LHS $\subseteq$ RHS is proved—as expected, given (9). If $\mathbb{Q}$ is subset closed, we get $(\dagger)\subseteq(*)$ as follows. Given q,t,u in $(\dagger)$, let $q':=t\cup u$. Then $t=\{\!\!\{\,b\,\}\!\!\}q'$ and $u=\{\!\!\{\,\neg b\,\}\!\!\}q$ because $\{\!\!\{\,b\,\}\!\!\}$ and $\{\!\!\{\,\neg b\,\}\!\!\}$ are filters. And $q'\in\mathbb{Q}$ by subset closure. Taking $q:=q'$ in $(*)$ completes the proof of RHS $\subseteq$ LHS. $\square$

Preservation of Subset Closure. In light of Lemma 3, we aim to restrict attention to h-transformers on subset closed sets. To this end we introduce a few notations. The subset-closed powerset operator $\breve{\wp}$ is defined on powersets $\wp A$, by

$$\mathbb{Q}\in\breve{\wp}(\wp A)\quad\text{iff}\quad\mathbb{Q}\subseteq\wp A\text{ and }\mathbb{Q}=\mathsf{ssc}\,\mathbb{Q}\text{ and }\mathbb{Q}\neq\emptyset\qquad(10)$$

To restrict attention to h-transformers of type $\breve{\wp}(\wp\Sigma) \to \breve{\wp}(\wp\Sigma)$ we must show that subset closure is preserved by the semantic constructs.

For any transformer φ, define $\mathsf{PSC}\,\varphi$ iff $(\supseteq ; \varphi) = (\varphi ; \supseteq)$. The acronym is explained by the lemma to follow. By definitions, the inclusion $(\supseteq ; \varphi) \supseteq (\varphi ; \supseteq)$ is equivalent to

$$\forall q, r \bullet \varphi q \supseteq r \Rightarrow \exists s \bullet q \supseteq s \land \varphi s = r \qquad (11)$$

Recall from Sect. 2 that the reverse, $(\supseteq ; \varphi) \subseteq (\varphi ; \supseteq)$, is monotonicity of φ.

Lemma 4. $\mathsf{PSC}\,\varphi$ *implies* $\langle\varphi\rangle$ *preserves subset closure.*

Proof. For any subset closed $\mathbb{Q}$, $\langle\varphi\rangle\mathbb{Q}$ is subset closed because $\langle\supseteq\rangle(\langle\varphi\rangle\mathbb{Q}) = \langle\varphi ; \supseteq\rangle\mathbb{Q} = \langle\supseteq ; \varphi\rangle\mathbb{Q} = \langle\varphi\rangle(\langle\supseteq\rangle\mathbb{Q}) = \langle\varphi\rangle\mathbb{Q}$ using functoriality of $\langle-\rangle$, $\mathsf{PSC}\,\varphi$, and $\mathsf{ssc}\,\mathbb{Q} = \mathbb{Q}$. $\qquad\square$

It is straightforward to show $\mathsf{PSC}\,\bot$. The following is a key fact, but also a disappointment that leads us away from nondeterminacy.

Lemma 5. *If R is a partial function (i.e., $R^{\cup} ; R \subseteq id$) then $\mathsf{PSC}\,\langle R\rangle$.*

Proof. In accord with (11) we show for any q, r that $r \subseteq \langle R\rangle q \Rightarrow \exists s \subseteq q \bullet \varphi s = r$. Suppose $r \subseteq \langle R\rangle q$. Let $s = (\langle R^{\cup}\rangle r) \cap q$, so for any x we have $x \in s$ iff $x \in q$ and $\exists y \in r \bullet xRy$. We have $s \subseteq q$ and it remains to show $\langle R\rangle s = r$, which holds because for any y

$$
\begin{array}{ll}
\quad y \in \langle R\rangle s \\
\equiv & \qquad\qquad\qquad\qquad\qquad \text{def } \langle-\rangle \\
\quad \exists x \bullet x \in s \land xRy \\
\equiv & \qquad\qquad\qquad\qquad\qquad \text{def } s \\
\quad \exists x \bullet x \in q \land (\exists z \bullet z \in r \land xRz) \land xRy \\
\equiv & \qquad\qquad\qquad\qquad\qquad R \text{ partial function} \\
\quad \exists x \bullet x \in q \land y \in r \land xRy \\
\equiv & \qquad\qquad\qquad\qquad\qquad \Leftarrow \text{ by } r \subseteq \langle R\rangle q \text{ and def } \langle-\rangle \\
\quad y \in r
\end{array}
$$

Using dots to show domain and range elements, the diagram on the left is an example R such that $\mathsf{PSC}\,\langle R\rangle$ but R is not a partial function. The diagram on the right is a relation, the image of which does not satisfy PSC.

As a consequence of Lemmas 4 and 5 we have the following.

Lemma 6. *If R is a partial function then $\langle\langle R\rangle\rangle : \wp^2 A \to \wp^2 B$ preserves subset closure.*

The Theorem. To prove $(\!| c |\!) = \langle\{\!| c |\!\}\rangle$, we want to identify a subset of $\wp\Sigma \to \wp\Sigma$ satisfying two criteria. First, $\{\!|-|\!\}$ can be defined within it, so in particular it is closed under G in Fig. 2. Second, on the subset, $\langle-\rangle$ is strict and continuous into $\breve{\wp}(\wp\Sigma) \to \breve{\wp}(\wp\Sigma)$, to enable the use of fixpoint fusion. Strictness is the reason[4] to disallow the empty set in (10); it makes $\bot$ (which equals $\langle\bot\rangle$) the least element, whereas otherwise the least element would be $\lambda Q \bullet \emptyset$. We need the subset to be closed under pointwise union, at least for chains, so that $\langle-\rangle$ is continuous.

Given that $\langle R \rangle$ is universally disjunctive for any R, Proposition 1 suggests restricting to universally disjunctive transformers. Lemma 4 suggests restricting to transformers satisfying PSC. But we were not able to show the universally disjunctive transformers satisfying PSC are closed under limits. We proceed as follows.

Define $\mathsf{Dom}\,\varphi = \{x \mid \varphi\{x\} \neq \emptyset\}$ and note that $\mathsf{Dom}\langle R \rangle = \mathrm{dom}R$ where $\mathrm{dom}R$ is the usual domain of a relation. By a straightforward proof we have:

Lemma 7. *For universally disjunctive φ and any r we have $\varphi r = \varphi(r \cap \mathsf{Dom}\,\varphi)$.*

Lemma 8. *For universally disjunctive φ, ψ with $\mathsf{Dom}\,\varphi \cap \mathsf{Dom}\,\psi = \emptyset$, if $\mathsf{PSC}\,\varphi$ and $\mathsf{PSC}\,\psi$ then $\mathsf{PSC}\,(\varphi \sqcup \psi)$.*

Proof. For any q, r with $r \subseteq (\varphi \sqcup \psi)q$ we need to show $\exists s \subseteq q \bullet (\varphi \sqcup \psi)s = r$. First observe

$$r \subseteq (\varphi \sqcup \psi)q$$
$$\equiv \quad \wr \text{ def } \sqcup \wr$$
$$r \subseteq \varphi q \cup \psi q$$
$$\equiv \quad \wr \text{ Lemma 7 } \wr$$
$$r \subseteq \varphi(q \cap \mathsf{Dom}\,\varphi) \cup \psi(q \cap \mathsf{Dom}\,\psi)$$
$$\Rightarrow \quad \wr \text{ set theory, letting } s = r \cap \varphi(q \cap \mathsf{Dom}\,\varphi) \text{ and } s' = r \cap \psi(q \cap \mathsf{Dom}\,\psi) \wr$$
$$r = s \cup s' \wedge s \subseteq \varphi(q \cap \mathsf{Dom}\,\varphi) \wedge s' \subseteq \psi(q \cap \mathsf{Dom}\,\psi)$$
$$\Rightarrow \quad \wr \text{ using } \mathsf{PSC}\,\varphi \text{ and } \mathsf{PSC}\,\psi \wr$$
$$\exists t, t' \bullet t \subseteq q \cap \mathsf{Dom}\,\varphi \wedge t' \subseteq q \cap \mathsf{Dom}\,\psi \wedge \varphi t = s \wedge \psi t' = s' \wedge s \cup s' = r \quad (*)$$

We use $(*)$ to show that $t \cup t'$ witnesses $\mathsf{PSC}\,(\varphi \sqcup \psi)$, as follows: $(\varphi \sqcup \psi)(t \cup t') = \varphi(t \cup t') \cup \psi(t \cup t') = \varphi t \cup \psi t' = s \cup s' = r$ using also the definition of $\sqcup$, and $\varphi(t \cup t') = \varphi t$ and $\psi(t \cup t') = \psi t'$ from Lemma 7 and $(*)$. $\qquad \square$

Lemma 9. *If Φ and Ψ preserve subset closure then $(\Phi \lhd b \rhd \Psi)Q$ is subset closed (regardless of whether Q is).*

Proof. Suppose q is in $(\Phi \lhd b \rhd \Psi)Q$ and $q' \subseteq q$. So according to (8) there are r, s, p with $p \in Q$, $q = r \cup s$, $r \in \Phi(\wp(\{\!| b |\!\}p))$, and $s \in \Psi(\wp(\{\!| \neg b |\!\}p))$. Let $r' = r \cap q'$ and $s' = s \cap q'$, so $r' \subseteq r$ and $s' \subseteq s$. Because powersets are subset closed, $\Phi(\wp(\{\!| b |\!\}p))$ and $\Psi(\wp(\{\!| \neg b |\!\}p))$ are subset closed, hence $r' \in \Phi(\wp(\{\!| b |\!\}p))$ and $s' \in \Psi(\wp(\{\!| \neg b |\!\}p))$. As $q' = r' \cup s'$, we have $q' \in (\Phi \lhd b \rhd \Psi)Q$. $\qquad \square$

---

[4] In [10], other reasons are given for using $\{\emptyset\}$ rather than $\emptyset$ as the false hyperproperty.

It is straightforward to prove that $\Phi \otimes \Psi$ preserves subset closure if Φ, Ψ do, similar to the proof of Lemma 9. By contrast, $\Phi \otimes \Psi$ does not preserve subset closure even if Φ and Ψ do.

Next, we confirm that $(\!|-\!|)$ can be defined within the monotonic functions $\breve{\wp}(\wp\Sigma) \to \breve{\wp}(\wp\Sigma)$.

Lemma 10. *For all $c, \mathbb{Q}$, if $\mathbb{Q}$ is subset closed then so is $(\!|c|\!)\mathbb{Q}$, provided that* PSC $\langle [\![atm]\!] \rangle$ *for every atm.*

Proof. By induction on c.

- atm: $(\!|atm|\!)$ is $\langle\langle [\![atm]\!] \rangle\rangle$ so by assumption PSC $\langle [\![atm]\!] \rangle$ and Lemma 1.
- skip: immediate.
- $c; d$: by definitions and induction hypothesis.
- $c \oplus d$: by induction hypothesis and observation above about $\otimes$.
- if b then c else d: by Lemma 9 and induction hypothesis.
- while b do c: Because $\bot$ is least in $\breve{\wp}(\wp\Sigma) \to \breve{\wp}(\wp\Sigma)$, we have $\bot \sqsubseteq \mathsf{H}\bot$, so using monotonicity of H we have Kleene iterates. Suppose $\mathbb{Q}$ is subset closed. To show lfp $\mathsf{H}\,\mathbb{Q}$ is subset closed, note that lfp $\mathsf{H} = \mathsf{H}^\gamma\,\mathbb{Q}$ where γ is some ordinal. We show that $\mathsf{H}^\alpha\,\mathbb{Q}$ is subset closed, for every α up to γ, by ordinal induction.
  - $\mathsf{H}^0\,\mathbb{Q} = \mathbb{Q}$ which is subset closed.
  - $\mathsf{H}^{\alpha+1}\,\mathbb{Q} = ((\!|c|\!)\,; \mathsf{H}^\alpha \triangleleft b \triangleright (\!|\mathsf{skip}|\!))\mathbb{Q}$ by definition of H. Now H^α preserves subset closure by the ordinal induction hypothesis, and $(\!|c|\!)$ preserves subset closure by the main induction hypothesis. So $(\!|c|\!)\,; \mathsf{H}^\alpha$ preserves subset closure, as does $(\!|\mathsf{skip}|\!)$. Hence $(\!|c|\!)\,; \mathsf{H}^\alpha \triangleleft b \triangleright (\!|\mathsf{skip}|\!)$ preserves subset closure by Lemma 9.
  - $\mathsf{H}^\beta\,\mathbb{Q} = (\sqcup_{\alpha<\beta}\mathsf{H}^\alpha)\mathbb{Q}$ (for non-0 limit ordinal β), which in turn equals $\cup_{\alpha<\beta}(\mathsf{H}^\alpha\,\mathbb{Q})$ because $\sqcup$ is pointwise. By induction, each $\mathsf{H}^\alpha\,\mathbb{Q}$ is subset closed, and closure is preserved by union, so we are done. □

Returning to the two criteria for a subset of $\wp\Sigma \to \wp\Sigma$, suppose $[\![atm]\!]$ is a partial function, for all atm—in short, **atoms are deterministic**. If in addition c is $\oplus$-free, then $[\![c]\!]$ is a partial function. Under these conditions, by Proposition 1, $\{\!|c|\!\}$ is the direct image of a partial function.

Let IPF be the subset of $\wp\Sigma \to \wp\Sigma$ that are direct images of partial functions, i.e., $\mathsf{IPF} = \{\varphi \in \wp\Sigma \to \wp\Sigma \mid \exists R \bullet \varphi = \langle R \rangle \text{ and } R^\cup\,; R \subseteq id\}$. Observe that IPF is closed under G, because for $\varphi \in \mathsf{IPF}$ with $\varphi = \langle R \rangle$ we have $\mathsf{G}\langle R \rangle = \langle [\![b]\!] \rangle\,; \langle [\![c]\!] \rangle\,; \langle R \rangle \sqcup \langle [\![\neg b]\!] \rangle = \langle [\![b]\!]\,; [\![c]\!]\,; R \rangle \sqcup \langle [\![\neg b]\!] \rangle = \langle [\![b]\!]\,; [\![c]\!]\,; R \cup [\![\neg b]\!] \rangle$ and the union is of partial functions with disjoint domains so it is a partial function. We have $\bot \sqsubseteq \mathsf{G}\bot$ because $\bot$ is the least element in $\wp\Sigma \to \wp\Sigma$. By Lemma 6, when $\langle - \rangle$ is restricted to IPF, its range is included in $\breve{\wp}(\wp\Sigma) \to \breve{\wp}(\wp\Sigma)$. In IPF, lubs of chains are given by pointwise union, so $\langle - \rangle$ is a strict and continuous function from IPF to $\breve{\wp}(\wp\Sigma) \to \breve{\wp}(\wp\Sigma)$.

To state the theorem, we write $\doteq$ for extensional equality on h-transformers of type $\breve{\wp}(\wp\Sigma) \to \breve{\wp}(\wp\Sigma)$, i.e., equal results on all subset closed $\mathbb{Q}$.

Theorem 1. $\langle\{\!|c|\!\}\rangle \doteq (\!|c|\!)$, *provided atoms are deterministic and c is $\oplus$-free.*

Proof. By induction on c. For the cases of skip, atoms, and ; the arguments preceding (6) are still valid. For conditional, observe

$$\langle\!\lbrace\, \text{if } b \text{ then } c \text{ else } d \,\rbrace\!\rangle$$
$$\doteq \qquad\qquad\qquad\qquad\qquad\qquad \text{semantics}$$
$$\langle\!\lbrace\, b \,\rbrace\!\rangle ; \lbrace\!\lbrace\, c \,\rbrace\!\rbrace \sqcup \lbrace\!\lbrace\, \neg b \,\rbrace\!\rbrace ; \lbrace\!\lbrace\, d \,\rbrace\!\rbrace\rangle$$
$$\doteq \qquad\qquad\qquad\qquad\qquad\qquad \text{Lemma 3}$$
$$\langle\!\lbrace\, c \,\rbrace\!\rangle \lhd b \rhd \langle\!\lbrace\, d \,\rbrace\!\rangle$$
$$\doteq \qquad\qquad\qquad\qquad\qquad\qquad \text{induction hypothesis}$$
$$(\!| c |\!) \lhd b \rhd (\!| d |\!)$$
$$\doteq \qquad\qquad\qquad\qquad\qquad\qquad \text{semantics}$$
$$(\!| \text{if } b \text{ then } c \text{ else } d |\!)$$

Finally, the loop:

$$\langle\!\lbrace\, \text{while } b \text{ do } c \,\rbrace\!\rangle$$
$$\doteq \qquad\qquad\qquad\qquad \text{semantics}$$
$$\langle\mathsf{lfp}\,\mathsf{G}\rangle$$
$$\doteq \qquad\qquad\qquad\qquad \text{fixpoint fusion, see below}$$
$$\mathsf{lfp}\,\mathsf{H}$$
$$\doteq \qquad\qquad\qquad\qquad \text{semantics}$$
$$(\!| \text{while } b \text{ do } c |\!)$$

The antecedent for fusion is $\forall\varphi \bullet \langle\mathsf{G}\varphi\rangle = \mathsf{H}\langle\varphi\rangle$ and it holds because for any φ:

$$\langle\mathsf{G}\varphi\rangle$$
$$\doteq \qquad\qquad\qquad\qquad\qquad\qquad \text{def G}$$
$$\langle\!\lbrace\, b \,\rbrace\!\rbrace ; \lbrace\!\lbrace\, c \,\rbrace\!\rbrace ; \varphi \sqcup \lbrace\!\lbrace\, \neg b \,\rbrace\!\rbrace\rangle$$
$$\doteq \qquad\qquad\qquad\qquad\qquad\qquad \text{skip law}$$
$$\langle\!\lbrace\, b \,\rbrace\!\rbrace ; \lbrace\!\lbrace\, c \,\rbrace\!\rbrace ; \varphi \sqcup \lbrace\!\lbrace\, \neg b \,\rbrace\!\rbrace ; \lbrace\!\lbrace\, \text{skip} \,\rbrace\!\rbrace\rangle$$
$$\doteq \qquad\qquad\qquad\qquad\qquad\qquad \text{Lemma 3}$$
$$\langle\!\lbrace\, c \,\rbrace\!\rbrace ; \varphi\rangle \lhd b \rhd \langle\!\lbrace\, \text{skip} \,\rbrace\!\rangle$$
$$\doteq \qquad\qquad\qquad\qquad\qquad\qquad \langle-\rangle \text{ distributes over } ;$$
$$\langle\!\lbrace\, c \,\rbrace\!\rangle ; \langle\varphi\rangle \lhd b \rhd \langle\!\lbrace\, \text{skip} \,\rbrace\!\rangle$$
$$\doteq \qquad\qquad\qquad\qquad\qquad\qquad \text{induction hypothesis}$$
$$(\!| c |\!) ; \langle\varphi\rangle \lhd b \rhd (\!| \text{skip} |\!)$$
$$\doteq \qquad\qquad\qquad\qquad\qquad\qquad \text{def H}$$
$$\mathsf{H}\langle\varphi\rangle$$

5   Specifications and Refinement

We wish to conceive of specifications as miraculous programs that can achieve by refusing to do, can choose the best angelically, and can compute the uncomputable. We wish to establish rigorous connections between programs and specifications, perhaps by deriving a program that can be automatically compiled for execution, perhaps by deriving a specification that can be inspected to determine the usefulness or trustworthiness of the program. A good theory may enable automatic derivation in one direction or the other, but should also account for ad hoc construction of proofs. Simple reasons should be expressed simply, so algebraic laws and transitive refinement chains are important. In this inconclusive

section, we return to the general notion of hyperproperty and consider how the
h-transformer semantics sheds light on refinement for hyperproperties. Initially
we leave aside the signature/semantics notations.

Let $R : \Sigma \multimap \Sigma$ be considered as a program, and $\mathbb{H}$ be a hyperproperty, that
is, $\mathbb{H}$ is a set of programs. Formally: $\mathbb{H} \in \wp(\Sigma \multimap \Sigma)$. For R to satisfy $\mathbb{H}$ means, by
definition, that $R \in \mathbb{H}$. The example of possibilistic noninterference shows that
in general trace refinement is unsound: $R \in \mathbb{H}$ does not follow from $S \in \mathbb{H}$ and
$S \supseteq R$. It does follow in the case that $\mathbb{H}$ is subset closed. Given that $\wp(\Sigma \multimap \Sigma)$
is a huge space, one may hope that specifications of practice interest may lie in
relatively tame subsets. Let us focus on subset closed hyperproperties, for which
one form of chain looks like

$$\mathbb{H} \ni S \supseteq \ldots \supseteq T \supseteq R \tag{12}$$

Although this is a sound way to prove $R \in \mathbb{H}$, it does not seem sufficient, at least
for examples like NI which require some degree of determinacy. The problem is
that for intermediate steps of trace refinement it is helpful to use nondeterminacy
for the sake of abstraction and underspecification, so finding suitable S and T
may be difficult. One approach to this problem is to use a more nuanced notion of
refinement, that preserves a hyperproperty of interest. For confidentiality, Banks
and Jacob explore this approach in the setting of UTP [6].

Another form of chain looks like

$$\mathbb{H} \supseteq \mathbb{S} \supseteq \ldots \supseteq \mathbb{T} \ni R \tag{13}$$

where most intermediate terms are at the hyper level, i.e., $\mathbb{S}$ and $\mathbb{T}$ are, like $\mathbb{H}$,
elements of $\wp(\Sigma \multimap \Sigma)$. The chain is a sound way to prove $R \in \mathbb{H}$, even if $\mathbb{H}$
is not subset closed. But in what way are the intermediates $\mathbb{S}, \mathbb{T}, \ldots$ expressed,
and by what reasoning are the containments established? What is the relevant
algebra, beyond elementary set theory?

The development in Sect. 4 is meant to suggest a third form of chain:

$$\ldots \sqsupseteq \mathbb{S} \sqsupseteq \ldots \sqsupseteq \mathbb{T} \sqsupseteq \langle\langle R \rangle\rangle \tag{14}$$

Here the intermediate terms are of type $\breve{\wp}(\wp\Sigma) \to \breve{\wp}(\wp\Sigma)$ and $\sqsubseteq$ is the pointwise
ordering (used already in Sect. 4). The good news is that if the intermediate
terms are expressed using program notations, they may be amenable to familiar
laws such as those of Kleene algebra with tests [23,38], for which relations are a
standard model. A corollary of Theorem 1 is that the laws hold for deterministic
terms expressed in the signature (4). To make this claim precise one might
spell out the healthiness conditions of elements in the range of $(\!|-|\!)$, but more
interesting would be to extend the language with specification constructs, using
(in)equational conditions like $R^{\cup} \, ; R \sqsubseteq id$ as antecedents in conditional laws of
healthy fragments. We leave this aside in order to focus on a gap in our story so
far.

The third form of chain is displayed with ellipses on the left because we
lack an account of specifications! Our leading example, NI, is defined as a set of

relations, whereas in (14) the displayed chain needs the specification, say Ψ, to have type $\breve{\wp}(\wp\Sigma) \to \breve{\wp}(\wp\Sigma)$. The closest we have come is the characterization $R \in \mathsf{NI}$ iff $\langle\langle R\rangle\rangle\mathbb{A} \subseteq \mathbb{A}$, see (5). But this is a set containment, whereas we seek Ψ with $R \in \mathsf{NI}$ iff $\Psi \sqsupseteq \langle\langle R\rangle\rangle$. In the rest of this section we sketch two ways to proceed.

On the face of it, $\Psi \sqsupseteq \langle\langle R\rangle\rangle$ seems problematic because $\sqsupseteq$ is an ordering on functions. Given a particular set $p \in \wp\Sigma$ with $\mathsf{Agrl}\,p$, a noninterfering R makes specific choice of value for lo whereas specification Ψ should allow any value for lo provided that the choice does not depend on the initial value for hi. One possibility is to escalate further and allow the specification to be a relation $\breve{\wp}(\wp\Sigma) \multimap \breve{\wp}(\wp\Sigma)$. To define such a relation, first lift the predicate Agrl on sets to the filter $\widehat{\mathsf{Agrl}} : \wp(\wp\Sigma) \multimap \wp(\wp\Sigma)$ defined by

$$\widehat{\mathsf{Agrl}}\,\mathbb{Q} = \{p \in \mathbb{Q} \mid \mathsf{Agrl}\,p\}$$

Note that $\mathbb{A} = \widehat{\mathsf{Agrl}}\,(\wp\Sigma)$. More to the point, $\widehat{\mathsf{Agrl}}\,\mathbb{Q} = \mathbb{Q}$ just if each $p \in \mathbb{Q}$ satisfies Agrl. Now define $\mathbf{NI}$, as a relation $\mathbf{NI} : \breve{\wp}(\wp\Sigma) \multimap \breve{\wp}(\wp\Sigma)$, by

$$\mathbb{P}\,\mathbf{NI}\,\mathbb{Q} \quad \text{iff} \quad \widehat{\mathsf{Agrl}}\,\mathbb{P} = \mathbb{P} \Rightarrow \widehat{\mathsf{Agrl}}\,\mathbb{Q} = \mathbb{Q}$$

This achieves the following: $R \in \mathsf{NI}$ iff $\mathbf{NI} \supseteq \langle\langle R\rangle\rangle$. But this inclusion does not compose transitively with $\sqsupseteq$ in the third form of chain, so we proceed no further in this direction.

The second way to proceed can be described using a variation on the h-transformer semantics of Sect. 4. It will lead us back to the second form of chain, (13), in particular for NI as $\mathbb{H}$. The idea resembles UTP models of reactive processes [19, Chap. 8], in which an event history is related to its possible extension. Here we use just pre-post traces, as follows. Let $\mathsf{Trc} = \Sigma \times \Sigma$. Consider a semantics $\{\!|-|\!\}'$ such that $\{\!|c|\!\}'$ has type $\wp\mathsf{Trc} \to \wp\mathsf{Trc}$. Instead of transforming an initial state to a final one (or rather, state set as in $\{\!|-|\!\}$), an initial trace (σ, τ) is mapped to traces (σ, v) for v with $\tau[\![c]\!]v$. The semantics $\{\!|-|\!\}'$ is not difficult to define (or see [3, sect. 2]). The upshot is that for $S \in \wp\mathsf{Trc}$, the trace set $\{\!|c|\!\}'S$ is the relation $S; [\![c]\!]$. In particular, let $init$ be id_Σ, viewed as an element of $\wp\mathsf{Trc}$. We get $\{\!|c|\!\}'init = [\![c]\!]$, the relation denoted by c. Lifting, we obtain a semantics $(\!|-|\!)'$, at the level $\wp^2\mathsf{Trc} \to \wp^2\mathsf{Trc}$, such that $(\!|c|\!)'\{init\}$ contains $[\![c]\!]$. (In light of Theorem 1 and the discussion preceding it, we do not expect $(\!|c|\!)'\{init\}$ to be just the singleton $\{[\![c]\!]\}$, as $\{init\}$ is not subset closed.) This suggests a chain of the form $\mathsf{NI} \supseteq \ldots \supseteq \Psi(\mathsf{ssc}\,\{init\}) \supseteq (\!|c|\!)'(\mathsf{ssc}\,\{init\}) \ni [\![c]\!]$ which proves that $[\![c]\!]$ satisfies NI—and which may be derived from a subsidiary chain of refinements like $\Psi \sqsupseteq (\!|c|\!)'$ as in the third form of chain, independent of the argument $\mathsf{ssc}\,\{init\}$.

This approach has been explored in the setting of abstract interpretation, where the intermediate terms are obtained as a computable approximation of a given program's semantics. To sketch the the idea we first review abstract interpretation for trace properties. Mathematically, abstract interpretation is very close to data refinement, where intermediate steps involve changes of data representation. For example, the state space Σ of $R : \Sigma \multimap \Sigma$ would be connected with another, say Δ, by a coupling relation $\rho : \Delta \multimap \Sigma$ subject to a simulation condition such as $S \,;\, \rho \supseteq \rho \,;\, R$, recall (3). With a functional coupling, the connection could be $\rho(\langle S \rangle \Delta) \sqsupseteq \langle R \rangle (\rho \Delta)$.

Let $T \in \Sigma \multimap \Sigma$ be a trace set intended as a trace property specification. In terms of the trace-computing semantics $\{\!|-|\!\}'$ above, c satisfies T provided that $T \supseteq \{\!| c |\!\}' init$. It can be proved by the following chain, ingredients of which are to be explained.

$$T \;\supseteq\; \gamma(\{\!| c |\!\}^{\sharp} a) \;\supseteq\; \{\!| c |\!\}'(\gamma\, a) \;\supseteq\; \{\!| c |\!\}' init$$

Here $\gamma : A \to \wp \mathsf{Trc}$ is like ρ above, mapping some convenient domain A to traces. The element $a \in A$ is supposed to be an approximation of the initial traces, i.e., $\gamma\, a \supseteq init$. Thus the containment $\{\!| c |\!\}'(\gamma\, a) \supseteq \{\!| c |\!\}' init$ is by monotonicity of semantics. The next containment, $\gamma(\{\!| c |\!\}^{\sharp} a) \supseteq \{\!| c |\!\}'(\gamma\, a)$, involves an "abstract" semantics $\{\!| c |\!\}^{\sharp} : A \to A$. Indeed, the containment is the soundness requirement for such semantics. What remains is the containment $T \supseteq \gamma(\{\!| c |\!\}^{\sharp}\, a)$ which needs to be checked somehow. Typically, γ is part of a Galois connection, i.e., it has a lower adjoint $\alpha : \wp \mathsf{Trc} \to A$ and the latter check is equivalent to $\alpha T \geq \{\!| c |\!\}^{\sharp}\, a$ where $\geq$ is the order on A. Ideally it is amenable to automation, but that is beside the point.

The point is to escalate this story to the hyper level, in a chain of this form:

$$\mathbb{H} \;\supseteq\; \gamma(\langle\!| c |\!\rangle^{\sharp} a) \;\supseteq\; \langle\!| c |\!\rangle'(\gamma\, a) \;\supseteq\; \langle\!| c |\!\rangle'(\mathsf{ssc}\,\{init\}) \;\ni\; \{\!| c |\!\}' init$$

Now γ has type $A \to \wp^2 \mathsf{Trc}$. Again, the abstract semantics should be sound—condition $\gamma(\langle\!| c |\!\rangle^{\sharp}\, a) \supseteq \langle\!| c |\!\rangle'(\gamma\, a)$—now with respect to a set-of-trace-set semantics $\langle\!|-|\!\rangle'$ that corresponds to our Fig. 3. The element $a \in A$ now approximates the set $\mathsf{ssc}\,\{init\}$ and $\langle\!| c |\!\rangle'(\gamma\, a) \supseteq \langle\!| c |\!\rangle'(\mathsf{ssc}\,\{init\})$ is by monotonicity. The step $\mathbb{H} \supseteq \gamma(\langle\!| c |\!\rangle^{\sharp}\, a)$ may again be checked at the abstract level as $\alpha\mathbb{H} \geq \langle\!| c |\!\rangle^{\sharp}\, a$. Of course $\mathbb{H}$ is a hyperproperty so the goal is to prove the program is an element of $\mathbb{H}$. This follows provided that $\langle\!| c |\!\rangle'(\mathsf{ssc}\,\{init\}) \ni \{\!| c |\!\}' init$, a connection like our Theorem 1 except for moving from $\wp \Sigma$ to $\wp \mathsf{Trc}$.

6 Related Work

The use of algebra in unifying theories of programming has been explored in many works including the book that led to the UTP meetings [19,20,37]. Methodologically oriented works include the books by Morgan [28] and by Bird and de Moor [8].

The term hyperproperty was introduced by Clarkson and Schneider who among other things mention that refinement at the level of trace properties is admissible for proving subset closed hyperproperties [10]. They point out that the topological classification of trace properties, i.e., safety and liveness, corresponds to similar notions dubbed hypersafety and hyperliveness. Subset closed hyperproperties strictly subsume hypersafety. As it happens, NI is in the class called 2-safety that specifies a property as holding for every pair of traces. For fixed k, one can encode k-safety by a product program, each trace of which represents k traces of the original. Some interesting requirements, such as quantitative information flow, are not in k-safety for any k.

Epistemic logic is the topic of a textbook [16] and has been explored in the security literature [5]. Mantel considers a range of security properties via closure operators [24]. The limited usefulness of trace refinement for proving NI even for deterministic programs, as in the chain (12), is discussed by Assaf and Pasqua [3,27]. The formulation of possibilistic noninterference as $\sim;R;\sim\ =\ R;\sim$ is due to Joshi and Leino [22] and resembles the formulation of Roscoe et al. [35].

The textbook of Back and von Wright [4] explores predicate transformer semantics and refinement calculus. The use of $sglt$ and $\ni$ in (1) is part of the extensive algebra connecting predicate transformers and relations using categorical notions [14]. Upward closed sets of predicates play an important role in that algebra [33], which should be explored in connection with the present investigation and its potential application to higher order programs. The extension of functional programming calculus to imperative refinement is one setting in which strong laws (Cartesian closure) for a well-behaved subset are expressed as implications with inequational antecedents [32,34]. These works use backward predicate transformers in order to model general specifications and in particular the combination of angelic and demonic nondeterminacy. Alternative models with similar aims can be found in Martin et al. [25] and Morris et al. [31].

A primary precursor to this paper is the dissertion work of Assaf, which targets refinement chains in the style of abstract interpretation [11,13]. Assaf's work [3] introduced a set-of-sets lifted semantics from which our h-transformer semantics is adapted. In keeping with the focus on static analysis, Assaf shows the lifted semantics is an approximation of the underlying one.[5] Assaf derives an abstraction $(\!|c|\!)^{\natural}$ for dependency from $(\!|c|\!)$ (for every c), by calculation, following Cousot [11] and similar to data refinement by calculation [17,30]. For this purpose and others, it is essential that loops be interpreted by fixpoint at the level of sets-of-sets, so standard fixpoint reasoning is applicable, as opposed to using $\langle\!\{$while b do $c\,\}\!\rangle$ which is not a fixed point per se. In fact Assaf derives two abstract semantics $(\!|c|\!)^{\natural}$: one for dependency (NI) and one that computes cardinality of low-variation, for quantitative information flow properties. The cardinality abstraction is not in k-safety for any k.

Pasqua and Mastroeni have aims similar to Assaf et al., and investigate several variations on set-of-set semantics of loops [27]. Our example in Sect. 4 is adapted from their work, which uses examples to suggest that Assaf's definition (called "mixed" in [27]) is preferable. They also point out that, for subset closed

---

[5] Assaf et al. use fixpoint fusion in the inequational form mentioned following (2), to prove soundness of the derived abstract semantics. Their inequational result corresponding to our Theorem is proved, in the loop case, using explicit induction on approximation chains. See the proof of Theorem 1 in [3].

hyperproperties, these variations are precise in the sense of our Theorem 1, strengthening the inequation in Assaf et al.[6]

A peculiarity of these works is the treatment of conditionals. Assaf et al. take the equation $(\!|$ if b then c else $d)\!| = \langle\{\!|$ if b then c else $d\}\!|\rangle$ as the definition of $(\!|$ if b then c else $d)\!|$, but this makes the definition of $(\!|-)\!|$ non-compositional and thus the inductive proofs a little sketchy. We do not discern an explicit definition in [27], but do find remarks like this: "The definition of the collecting hypersemantics is just the additive lift... for every statement, except for the while case." As a description of fact, it is true for subset closed hyperproperties (see our Theorem 1). But it is unsuitable as a definition. We show that a proper definition is possible.

Non-compositionality for sequence and conditional seems difficult to reconcile with proofs by induction on program structure. It also results in anomalies, e.g., the formulation in Assaf et al. means that in case c is a loop, the semantics of c is different from the semantics of if $true$ then c else skip. The obscurity is rectified when the semantics is restricted to subset closed sets, as spelled out in detail in an unpublished note [15], confirming remarks by Pasqua and Mastroeni [26] elaborated in [27] and in the thesis of Pasqua.

Although similar to results in the preceding work, our results are novel in a couple of ways. Our semantics is for a language with nondeterminacy, unlike theirs. Of course, nondeterministic programs typically fail to satisfy NI and related properties, and our theorem is restricted to deterministic programs. A minor difference is that they formulate loop semantics in a standard form mentioned in Footnote 3 that asserts the negated guard following the fixpoint rather than as part of it. Those works use semantics mapping traces to traces (or sets of sets thereof), like our $\{\!|-\}\!|'$ and $(\!|-)\!|'$, said to be needed in order to express dependency. We have shown that states-to-states is sufficient to exhibit both the anomaly and its resolution. It suffices for specifications of the form (5) and may facilitate further investigation owing to its similarity to many variations on relational and transformer semantics.

Apropos NI, the formulation (5) is robust in the sense that it generalizes to more nuanced notions of dependency: $\langle\langle R\rangle\rangle\mathbb{P} \subseteq \mathbb{Q}$ where $\mathbb{P}$ expresses agreement on some projections of the input (e.g., agreement on whether a password guess is correct, or agreement on some aggregate value derived from a sensitive database) and $\mathbb{Q}$ expresses agreement on the observable output values. Such policies are the subject of [36].

Banks and Jacob [6] formalize general confidentiality policies in UTP and introduce a family of confidentiality preserving refinement relations. The ideas are developed further in subsequent work where confidentiality-violating refinements are represented as miracles [7] and knowledge is explicitly represented by sets encoding alternate executions, an idea that has appeared in other guises [2, 29].

---

[6] Displayed formula $\{\{\!|c\}\!|T \mid T \in \mathbb{T}\} \subseteq (\!|c)\!|\mathbb{T}$ following Theorem 1 of [3].

7   Conclusion

We have given what, to the best of our knowledge, is the first compositional definition of semantics at the hyper level. Moreover, we proved that it is the lift of a standard semantics when restricted to subset closed hyperproperties. The latter is a "forward collecting semantics" in the terminology of abstract interpretation. The new semantics includes nondeterministic constructs, although the lifting equivalence is only proved for the deterministic fragment.

Although deterministic noninterference is a motivating example, there are other interesting hyperproperties such as quantitative information flow that can be expressed in subset closed form and which are meaningful for nondeterministic programs. This is one motivation for further investigation including the following questions. • Restricting to subset closed hyperproperties is sufficient to make possible a compositional fixpoint semantics at the hyper level that accurately represents the underlying semantics—is it necessary? • The h-transformer semantics allows nondeterministic choice and nondeterministic atoms that satisfy PSC (in light of Lemma 10), and in fact the definitions can be used for a semantics in $\wp(\wp\Sigma) \to \wp(\wp\Sigma)$, into which $\breve{\wp}(\wp\Sigma) \to \breve{\wp}(\wp\Sigma)$ embeds nicely owing to joins being pointwise— but what exactly is the significance of determinacy? • Are disjunctive transformers satisfying PSC closed under join? What is a good characterization of transformers that are images of relations?

A person not familiar with unifying theories of programming may wonder whether programs are specifications. Indeed, the author was once criticized by a famous computer scientist who objected to refinement calculi on the grounds that underspecification and nondeterminacy are distinct notions that ought not be confused—though years later he published a soundness proof for a program logic, in which that confusion is exploited to good effect. A positive answer to the question can be justified by embedding programs in a larger space of specifications, in a way that faithfully reflects a given semantics of programs. Our theorem is a result of this kind. The larger space makes it possible to express important requirements such as noninterference. However, we do not put forward a compelling notion of specification that encompasses hyperproperties and supports a notion of refinement analogous to existing notions for trace properties. Rather, we hope the paper inspires or annoys the reader enough to provoke further research.

Acknowledgements. Anonymous reviewers offered helpful suggestions and pointed out errors, omissions, and infelicities in an earlier version.

The authors were partially supported by NSF award 1718713.

References

1. Aarts, C., et al.: Fixed-point calculus. Inf. Process. Lett. **53**(3), 131–136 (1995)
2. Assaf, M., Naumann, D.A.: Calculational design of information flow monitors. In: Computer Security Foundations (2016)

3. Assaf, M., Naumann, D.A., Signoles, J., Totel, É., Tronel, F.: Hypercollecting semantics and its application to static analysis of information flow. In: POPL (2017)
4. Back, R.J., von Wright, J.: Refinement Calculus: A Systematic Introduction. Springer, New York (1998). https://doi.org/10.1007/978-1-4612-1674-2
5. Balliu, M., Dam, M., Guernic, G.L.: Epistemic temporal logic for information flow security. In: Programming Languages and Analysis for Security (2011)
6. Banks, M.J., Jacob, J.L.: Unifying theories of confidentiality. In: Qin, S. (ed.) UTP 2010. LNCS, vol. 6445, pp. 120–136. Springer, Heidelberg (2010). https://doi.org/10.1007/978-3-642-16690-7_5
7. Banks, M.J., Jacob, J.L.: On integrating confidentiality and functionality in a formal method. Formal Aspects Comput. **26**(5), 963–992 (2014)
8. Bird, R., de Moor, O.: Algebra of Programming. Prentice-Hall, Upper Saddle River (1996)
9. Clarkson, M.R., Finkbeiner, B., Koleini, M., Micinski, K.K., Rabe, M.N., Sánchez, C.: Temporal logics for hyperproperties. In: Abadi, M., Kremer, S. (eds.) POST 2014. LNCS, vol. 8414, pp. 265–284. Springer, Heidelberg (2014). https://doi.org/10.1007/978-3-642-54792-8_15
10. Clarkson, M.R., Schneider, F.B.: Hyperproperties. J. Comput. Secur. **18**(6), 1157–1210 (2010)
11. Cousot, P.: The calculational design of a generic abstract interpreter. In: Broy, M., Steinbrüggen, R. (eds.) Calculational System Design. NATO ASI Series F, vol. 173. IOS Press, Amsterdam (1999)
12. Cousot, P.: Constructive design of a hierarchy of semantics of a transition system by abstract interpretation. Theor. Comput. Sci. **277**(1–2), 47–103 (2002)
13. Cousot, P., Cousot, R.: Systematic design of program analysis frameworks. In: POPL (1979)
14. Gardiner, P.H., Martin, C.E., de Moor, O.: An algebraic construction of predicate transformers. Sci. Comput. Program. **22**, 21–44 (1994)
15. Gotliboym, M., Naumann, D.A.: Some observations on hypercollecting semantics and subset closed hyperproperties. https://www.cs.stevens.edu/~naumann/pub/noteSSC.pdf
16. Halpern, J.Y., Fagin, R., Moses, Y., Vardi, M.Y.: Reasoning About Knowledge. MIT Press, Cambridge (1995)
17. He, J., Hoare, C.A.R., Sanders, J.W.: Data refinement refined resume. In: Robinet, B., Wilhelm, R. (eds.) ESOP 1986. LNCS, vol. 213, pp. 187–196. Springer, Heidelberg (1986). https://doi.org/10.1007/3-540-16442-1_14
18. Hoare, C.A.R., Lauer, P.E.: Consistent and complementary formal theories of the semantics of programming languages. Acta Inf. **3**, 135–153 (1974)
19. Hoare, C., He, J.: Unifying Theories of Programming. Prentice-Hall, Upper Saddle River (1998)
20. Hoare, T., Möller, B., Struth, G., Wehrman, I.: Concurrent Kleene algebra and its foundations. J. Log. Algebr. Program. **80**(6), 266–296 (2011)
21. Jacob, J.: Security specifications. In: IEEE Symposium on Security and Privacy (1988)
22. Joshi, R., Leino, K.R.M.: A semantic approach to secure information flow. Sci. Comput. Program. **37**(1–3), 113–138 (2000)
23. Kozen, D.: On Hoare logic and Kleene algebra with tests. ACM Trans. Comput. Log. **1**(1), 60–76 (2000)
24. Mantel, H.: On the composition of secure systems. In: IEEE Symposium on Security and Privacy (2002)

25. Martin, C.E., Curtis, S.A., Rewitzky, I.: Modelling angelic and demonic nondeterminism with multirelations. Sci. Comput. Program. **65**(2), 140–158 (2007)
26. Mastroeni, I., Pasqua, M.: Hyperhierarchy of semantics - a formal framework for hyperproperties verification. In: Ranzato, F. (ed.) SAS 2017. LNCS, vol. 10422, pp. 232–252. Springer, Cham (2017). https://doi.org/10.1007/978-3-319-66706-5_12
27. Mastroeni, I., Pasqua, M.: Verifying bounded subset-closed hyperproperties. In: Podelski, A. (ed.) SAS 2018. LNCS, vol. 11002, pp. 263–283. Springer, Cham (2018). https://doi.org/10.1007/978-3-319-99725-4_17
28. Morgan, C.: Programming from Specifications, 2nd edn. Prentice Hall, Upper Saddle River (1994)
29. Morgan, C.: The shadow knows: refinement and security in sequential programs. Sci. Comput. Program. **74**(8), 629–653 (2009)
30. Morgan, C., Gardiner, P.: Data refinement by calculation. Acta Inf. **27**, 481–503 (1990)
31. Morris, J.M., Bunkenburg, A., Tyrrell, M.: Term transformers: a new approach to state. ACM Trans. Program. Lang. Syst. **31**(4), 16 (2009)
32. Naumann, D.A.: Data refinement, call by value, and higher order programs. Formal Aspects Comput. **7**, 652–662 (1995)
33. Naumann, D.A.: A categorical model for higher order imperative programming. Math. Struct. Comput. Sci. **8**(4), 351–399 (1998)
34. Naumann, D.A.: Towards patterns for heaps and imperative lambdas. J. Log. Algebraic Methods Program. **85**(5), 1038–1056 (2016)
35. Roscoe, A.W., Woodcock, J.C.P., Wulf, L.: Non-interference through determinism. In: Gollmann, D. (ed.) ESORICS 1994. LNCS, vol. 875, pp. 31–53. Springer, Heidelberg (1994). https://doi.org/10.1007/3-540-58618-0_55
36. Sabelfeld, A., Sands, D.: Dimensions and principles of declassification. J. Comput. Secur. **17**, 517–548 (2007)
37. Sampaio, A.: An Algebraic Approach to Compiler Design. AMAST Series in Computing, vol. 4. World Scientific, Singapore (1997)
38. Struth, G.: On the expressive power of Kleene algebra with domain. Inf. Process. Lett. **116**(4), 284–288 (2016)

Connecting Fixpoints of Computations with Strict Progress

Walter Guttmann(✉)

Department of Computer Science and Software Engineering,
University of Canterbury, Christchurch, New Zealand
walter.guttmann@canterbury.ac.nz

Abstract. We study the semantics of recursion for computations that make strict progress. The underlying unified computation model has an abstract notion of progress, which instantiates in ways such as longer traces, passing of real time, or counting the number of steps. Recursion is given by least fixpoints in a unified approximation order. Other time-based models define the semantics of recursion by greatest fixpoints in the implication order. We give sufficient criteria for when least fixpoints in the approximation order coincide with greatest fixpoints in the implication order.

1 Introduction

A recursive computation has the form $x = f(x)$, where the function f specifies what happens in one unfolding step of the recursion in terms of x, which captures recursive invocations. For example, unfolding the loop while b do a results in the recursion while b do a = if b then a; while b do a else skip. Hence while b do a is a solution of the equation x = if b then a; x else skip using a suitable semantics for the conditional. Solutions of the equation $x = f(x)$ are fixpoints of the function f. It is therefore not surprising that fixpoints have been widely used to define the semantics of recursive computations [4,5,12,16,32,37,39].

In general, a function may have several fixpoints. For example, every x is a fixpoint of the identity function. The corresponding recursion equation $x = x$ may seem contrived, but it gives the semantics of the endless loop while true do skip since if true then skip; x else skip = skip; $x = x$ in many computation models. In this and similar cases, the question arises which fixpoint to choose as the semantics of recursion. Different computation models give different answers.

A common solution is to define an order on the computations and choose the least or greatest fixpoint in this order. Two orders are relevant for computation models: the implication order and the approximation order [5]. Conceptually they serve different purposes. The implication order establishes when a program implements a specification and is typically concerned with the amount of non-determinism exhibited by computations. The approximation order deals with the semantics of recursion; successive unfoldings of a recursion yield better approximations to its semantics, which emerges as the limit. Confusion sometimes occurs

© Springer Nature Switzerland AG 2019
P. Ribeiro and A. Sampaio (Eds.): UTP 2019, LNCS 11885, pp. 62–79, 2019.
https://doi.org/10.1007/978-3-030-31038-7_4

because in some computation models the approximation order happens to coincide with the implication order or its converse (the refinement order). For example, in the Unifying Theories of Programming (UTP) recursion is defined by least fixpoints in the refinement order [28]. In general, however, a better approximation is not tied to either a decrease or an increase in non-determinism.

Justification of a particular choice of fixpoints for a proposed computation model is usually left implicit (though explicit connections have sometimes been made [8]). Specifically, we note that apparently different choices are made for computation models with different notions of progress. A dedicated approximation order (different from implication) has been used in Boolean-time models, which are models that distinguish between termination and non-termination but do not have a finer notion of elapsing time [4,5]. Greatest fixpoints with respect to implication have been used in models that measure time using natural numbers or real numbers [24].

The aim of this paper is to provide additional confidence that the choices of fixpoints made for several computation models make sense. To this end, we study the relationship between the greatest fixpoint in the implication order and the least fixpoint in the approximation order. We do this in a unified computation model that supports an abstract notion of progress, developed in previous work [21]. Computations in this model are sequential and may be non-deterministic; single executions may be finite, aborting, incomplete or infinite. Progress can be measured in terms of Boolean-time, abstract time, real time, traces and other instances. The model supports a unified approximation order in addition to the implication order. It can therefore be used to investigate how extremal fixpoints in these orders are connected.

We give sufficient conditions for when least fixpoints in the approximation order coincide with greatest fixpoints in the implication order. We interpret these conditions for various notions of progress. The findings of this paper can be summarised as follows. The availability of real time in the model is not sufficient for the fixpoints to coincide even if strict progress is assumed, that is, if time cannot stand still. However, the two studied fixpoints coincide if a uniform lower bound can be given for the progress made in each unfolding of the recursion.

Knowing circumstances in which the implication order can be used to define recursion is helpful since it is arguably simpler than the approximation order. Moreover the implication order is fundamental for reasoning about the correctness of computations, so a direct connection with the approximation order facilitates this task.

Section 2 describes the setup for our study including basic algebraic structures and the unified computation model. The following sections comprise new results, which form the contributions of this paper. We outline the overall strategy in Sect. 3. Section 4 contains results that can be derived algebraically, based on complete lattices, semirings and an iteration operation. In particular, we prove upper bounds for greatest fixpoints in the implication order by capturing the excess over corresponding least fixpoints in the same order. Section 5 contains results that are specific to the unified computation model referring to various notions of progress. In particular, we give further bounds on the excess so that the greatest fixpoint in the implication order coincides with the least fixpoint in the approximation order. We discuss what the conditions that enable these bounds mean for different kinds of progress.

2   Basic Definitions

This section presents the algebraic structures and the unified computation model used in the remainder of the paper. The discussion is based on [21].

2.1   Algebraic Structures

A *bounded distributive lattice* is an algebraic structure $(S, \sqcup, \sqcap, \bot, \top)$ with the following axioms:

$$x \sqcup (y \sqcup z) = (x \sqcup y) \sqcup z \qquad\qquad x \sqcap (y \sqcap z) = (x \sqcap y) \sqcap z$$
$$x \sqcup y = y \sqcup x \qquad\qquad x \sqcap y = y \sqcap x$$
$$x \sqcup x = x \qquad\qquad x \sqcap x = x$$
$$\bot \sqcup x = x \qquad\qquad \top \sqcap x = x$$
$$x \sqcup (y \sqcap z) = (x \sqcup y) \sqcap (x \sqcup z) \qquad x \sqcap (y \sqcup z) = (x \sqcap y) \sqcup (x \sqcap z)$$
$$x \sqcup (x \sqcap y) = x \qquad\qquad x \sqcap (x \sqcup y) = x$$

The *lattice order* $x \sqsubseteq y \Leftrightarrow x \sqcup y = y \Leftrightarrow x \sqcap y = x$ has least element $\bot$, greatest element $\top$, least upper bound operation $\sqcup$ and greatest lower bound operation $\sqcap$. The operations $\sqcup$ and $\sqcap$ are $\sqsubseteq$-isotone.

A bounded distributive lattice S is *MID-complete* if each $A \subseteq S$ has a meet $\sqcap A$ with the following axioms, including meet-infinite distributivity (MID):

$$\forall a \in A : \textstyle\bigsqcap A \sqsubseteq a \qquad (\forall a \in A : x \sqsubseteq a) \Rightarrow x \sqsubseteq \textstyle\bigsqcap A \qquad x \sqcup \textstyle\bigsqcap A = \textstyle\bigsqcap_{a \in A} x \sqcup a$$

We use $\bigsqcap_{a \in A} f(a) = \bigsqcap \{f(a) \mid a \in A\}$ and similar abbreviations.

An *idempotent semiring* without a right annihilator is an algebraic structure $(S, \sqcup, \cdot, \bot, 1)$ with the following axioms:

$$x \sqcup (y \sqcup z) = (x \sqcup y) \sqcup z \quad x \cdot (y \cdot z) = (x \cdot y) \cdot z \quad x \cdot (y \sqcup z) = (x \cdot y) \sqcup (x \cdot z)$$
$$x \sqcup y = y \sqcup x \qquad\qquad 1 \cdot x = x \qquad\qquad (x \sqcup y) \cdot z = (x \cdot z) \sqcup (y \cdot z)$$
$$x \sqcup x = x \qquad\qquad x \cdot 1 = x \qquad\qquad \bot \cdot x = \bot$$
$$\bot \sqcup x = x$$

Note that $x \cdot \bot = \bot$ is not an axiom. The operation $\cdot$ is $\sqsubseteq$-isotone. We assume the operation $\cdot$ has higher precedence than $\sqcup$ and $\sqcap$ and abbreviate $x \cdot y$ as xy. Powers in semirings are defined by $x^0 = 1$ and $x^{i+1} = xx^i$ for $x \in S$ and $i \in \mathbb{N}$. It follows that $x^{i+1} = x^i x$.

A *lattice-ordered semiring* is an algebraic structure $(S, \sqcup, \cdot, \sqcap, \bot, 1, \top)$ such that $(S, \sqcup, \sqcap, \bot, \top)$ is a bounded distributive lattice and $(S, \sqcup, \cdot, \bot, 1)$ is an idempotent semiring without a right annihilator.

A *lattice-ordered itering* adds to a lattice-ordered semiring a unary operation $^\circ$ with higher precedence than $\cdot$ and satisfying the sumstar and productstar equations of [7] and two simulation axioms [17]:

$$(x \sqcup y)^\circ = (x^\circ y)^\circ x^\circ \qquad zx \sqsubseteq yy^\circ z \sqcup w \Rightarrow zx^\circ \sqsubseteq y^\circ (z \sqcup wx^\circ)$$
$$(xy)^\circ = 1 \sqcup x(yx)^\circ y \qquad xz \sqsubseteq zy^\circ \sqcup w \Rightarrow x^\circ z \sqsubseteq (z \sqcup x^\circ w)y^\circ$$

The operation $^\circ$ is $\sqsubseteq$-isotone. Moreover the unfold property $x^\circ = 1 \sqcup xx^\circ$ and the related sumstar property $(x \sqcup y)^\circ = x^\circ(yx^\circ)^\circ$ follow. The itering operation generalises the Kleene star but has many other instances, for example, in omega algebras [6] and demonic refinement algebras [40]. In order to cover these instances, the simulation axioms of iterings generalise simpler simulation properties $zx \sqsubseteq yz \Rightarrow zx^\circ \sqsubseteq y^\circ z$ and $xz \sqsubseteq zy \Rightarrow x^\circ z \sqsubseteq zy^\circ$ suggested in [7] for Kleene algebras [31] and with applications in program transformation and data refinement [3, 6].

In many computation models, the operation $\sqcup$ represents non-deterministic choice, the operation $\cdot$ sequential composition, the operation $\sqcap$ conjunction, $\bot$ the computation with no executions, 1 the computation that does not change the state, $\top$ the computation with all executions, and $\sqsubseteq$ the implication order.

In particular, UTP designs [28] form instances of the above algebras, as do prescriptions, extended designs, conscriptions, extended conscriptions and other variants of designs discussed in the literature [10, 11, 18, 23]. Further works modelling computations based on lattices, semirings and related algebras include [2, 3, 6, 22, 29, 31, 33, 36, 40].

A (binary, homogeneous) *relation* $R : A \leftrightarrow A$ on a set $A \neq \emptyset$ is a set $R \subseteq A \times A$. Important constants are the empty relation $O = \emptyset$, the identity relation $I = \{(x, x) \mid x \in A\}$ and the universal relation $\top = A \times A$. The composition of relations Q and R is defined by $Q \cdot R = \{(x, z) \mid \exists y : (x, y) \in Q \wedge (y, z) \in R\}$. Relations on A form a lattice-ordered semiring $(2^{A \times A}, \cup, \cdot, \cap, O, I, \top)$, which is MID-complete and satisfies right annihilation $QO = O$. See [35] for further details about relations.

These definitions generalise to heterogeneous relations $R : A \leftrightarrow B$, which are sets $R \subseteq A \times B$ where A and B may differ. Heterogeneous relations form categories and their operations apply to arguments of suitable types [14, 34].

Let S be a set partially ordered by $\sqsubseteq$ and let $f : S \to S$. Provided they exist, the $\sqsubseteq$-least and $\sqsubseteq$-greatest *fixpoints* of f are denoted by μf and νf, respectively:

$$f(\mu f) = \mu f \qquad f(x) = x \Rightarrow \mu f \sqsubseteq x$$
$$f(\nu f) = \nu f \qquad f(x) = x \Rightarrow \nu f \sqsupseteq x$$

For functions $f, g : S \to S$ such that g is $\sqsubseteq$-isotone and the involved fixpoints exist, the two rolling properties $\mu(g \circ f) = g(\mu(f \circ g))$ and $\nu(g \circ f) = g(\nu(f \circ g))$ follow [1, 9]. If S is a complete lattice and f is $\sqsubseteq$-isotone, both μf and νf exist by Tarski's fixpoint theorem [38].

2.2   A Unified Model for Computations with Progress

We discuss a model that represents computations by four relations and describes different notions of progress in a uniform way. Computations are sequential and may be non-deterministic; a computation comprises single executions, each of which may be finite, aborting, incomplete or infinite. We explain this model to the extent necessary for this paper; for further details and related work see [21].

The state space A of a computation is partitioned into two sets A_{fin} and A_∞ comprising the finite and infinite parts. The following are examples of such a separation, where D is the set of values that program variables can take:

A_{fin}	A_∞	model
D	$\{\infty\}$	Boolean time; computation terminates or does not terminate
$D \times \mathbb{N}$	$\{\infty\}$	abstract time; steps are counted
$D \times \mathbb{R}$	$\{\infty\}$	real time; a clock is used
D^+	D^ω	traces; finite and infinite sequences over D

Behavioural aspects of computations are modelled by relations on the state space. Two relations $\mathsf{F} : A_{\text{fin}} \leftrightarrow A_{\text{fin}}$ and $\mathsf{F}_\infty : A_{\text{fin}} \leftrightarrow A_\infty$ describe finite progress and progress from the finite to the infinite. We assume that F is a preorder, that is, $\mathsf{I} \subseteq \mathsf{F} = \mathsf{F}^2$, and that the related transitivity property $\mathsf{FF}_\infty = \mathsf{F}_\infty$ holds. For the above examples, these constants are:

$\mathsf{F} : A_{\text{fin}} \leftrightarrow A_{\text{fin}}$	A_{fin}	$\mathsf{F}_\infty : A_{\text{fin}} \leftrightarrow A_\infty$	A_∞	model
T	D	T	$\{\infty\}$	Boolean time
$\{((v,t),(v',t')) \mid t \leq t'\}$	$D \times \mathbb{N}$	T	$\{\infty\}$	abstract time
$\{((v,t),(v',t')) \mid t \leq t'\}$	$D \times \mathbb{R}$	T	$\{\infty\}$	real time
$\preceq$	D^+	$\preceq$	D^ω	traces

Traces use the prefix relation $\preceq$ on finite and infinite sequences.

In a computation, four relations N, P, Q and R are used to describe different kinds of execution. The relation R represents the finite executions, which terminate successfully. The relation P represents executions that abort due to an error. The relation N represents incomplete executions, which are unproductive and used in approximation. The relation Q represents actually infinite executions.

Formally, a computation $(N|P|Q|R)$ comprises relations $N, P, R : A_{\text{fin}} \leftrightarrow A_{\text{fin}}$ and $Q : A_{\text{fin}} \leftrightarrow A_\infty$ satisfying the *progress requirements* $N, P, R \subseteq \mathsf{F}$ and $Q \subseteq \mathsf{F}_\infty$ and the *closure requirements* $N\mathsf{F} \subseteq N$ and $N\mathsf{F}_\infty \subseteq Q$. It is defined by the following matrix:

$$(N|P|Q|R) = \begin{pmatrix} \mathsf{I} & \mathsf{O} & \mathsf{O} & \mathsf{O} \\ \mathsf{O} & \mathsf{I} & \mathsf{O} & \mathsf{O} \\ \mathsf{O} & \mathsf{O} & \mathsf{I} & \mathsf{O} \\ N & P & Q & R \end{pmatrix}$$

In the trace model, the progress requirements correspond to UTP's healthiness condition R1, which specifies that traces can only get longer [28]. More details about this and the progress and closure requirements are given in [21].

The matrix definition of a computation helps to derive basic operations, which elaborate as follows.

- Non-deterministic choice is the componentwise union of the matrices:

$$(N_1|P_1|Q_1|R_1) \sqcup (N_2|P_2|Q_2|R_2) = (N_1 \cup N_2|P_1 \cup P_2|Q_1 \cup Q_2|R_1 \cup R_2)$$

- Sequential composition is given by the matrix product, where union and relational composition replace addition and multiplication:

$$(N_1|P_1|Q_1|R_1) \cdot (N_2|P_2|Q_2|R_2) = (N_1 \cup R_1 N_2|P_1 \cup R_1 P_2|Q_1 \cup R_1 Q_2|R_1 R_2)$$

- Conjunction is the componentwise intersection of the matrices:

$$(N_1|P_1|Q_1|R_1) \sqcap (N_2|P_2|Q_2|R_2) = (N_1 \cap N_2|P_1 \cap P_2|Q_1 \cap Q_2|R_1 \cap R_2)$$

- The implication order is the componentwise set inclusion order:

$$(N_1|P_1|Q_1|R_1) \sqsubseteq (N_2|P_2|Q_2|R_2) \Leftrightarrow N_1 \subseteq N_2 \wedge P_1 \subseteq P_2 \wedge Q_1 \subseteq Q_2 \wedge R_1 \subseteq R_2$$

- The computation with no executions is

$$\bot = (0|0|0|0)$$

- The computation that does not change the state is

$$1 = (0|0|0|\mathsf{I})$$

- The computation with all executions is

$$\top = (\mathsf{F}|\mathsf{F}|\mathsf{F}_\infty|\mathsf{F})$$

- The computation with all incomplete and infinite executions is

$$\mathsf{L} = (\mathsf{F}|0|\mathsf{F}_\infty|0)$$

- The approximation order is:

$$(N_1|P_1|Q_1|R_1) \sqsubseteq\!\!\!\!\!\! = (N_2|P_2|Q_2|R_2) \Leftrightarrow N_2 \subseteq N_1 \wedge P_1 \subseteq P_2 \subseteq P_1 \cup N_1 \wedge$$
$$Q_2 \subseteq Q_1 \wedge R_1 \subseteq R_2 \subseteq R_1 \cup N_1$$

With these operations, the set S of computations forms a lattice-ordered semiring $(S, \sqcup, \cdot, \sqcap, \bot, 1, \top)$, which has the lattice order $\sqsubseteq$ and is MID-complete. This can be extended to a lattice-ordered itering; details including the definition of $^\circ$ for computations are given in [21].

As usual, the implication order reflects the amount of non-determinism of computations and the notion of a computation refining another if the former contains a subset of the executions of the latter. The four subset relationships $N_1 \subseteq N_2$ and $P_1 \subseteq P_2$ and $Q_1 \subseteq Q_2$ and $R_1 \subseteq R_2$ state this for the four kinds of execution represented in our model.

The above approximation order has been derived in [21] using algebraic techniques. Intuitively it states that a more precise approximation can add finite and aborting executions ($P_1 \subseteq P_2 \wedge R_1 \subseteq R_2$) provided they extend incomplete executions ($P_2 \subseteq P_1 \cup N_1 \wedge R_2 \subseteq R_1 \cup N_1$), and can remove only incomplete and infinite executions ($N_2 \subseteq N_1 \wedge Q_2 \subseteq Q_1$).

Let S be the computations in our model and let $f : S \to S$. Provided it exists, the $\sqsubseteq$-least fixpoint of f is denoted by κf:

$$f(\kappa f) = \kappa f \qquad\qquad f(x) = x \Rightarrow \kappa f \sqsubseteq x$$

The following result is a consequence of [21, Theorems 2 and 4]. Item 4 uses the abbreviation $c(x) = n(\mathsf{L})\top \sqcap x$ and the operation n, which elaborates to $n(N|P|Q|R) = (\mathsf{O}|\mathsf{O}|Q|N)$ in our computation model.

Proposition 1. *Let S be the computations in our model.*

1. *The relation $\sqsubseteq$ is a partial order with least element L.*
2. *The operations $\sqcup$ and $\cdot$ and $^\circ$ are $\sqsubseteq$-isotone.*

Let $f : S \to S$ be $\sqsubseteq$- and $\sqsubseteq$-isotone. Then the following are equivalent:

3. *κf exists and $\kappa f = (\nu f \sqcap \mathsf{L}) \sqcup \mu f$.*
4. *$c(\nu f) \sqsubseteq (\nu f \sqcap \mathsf{L}) \sqcup \mu f \sqcup n(\nu f)\top$.*

Items 1 and 2 state basic properties of the approximation order and operations used for defining program constructs. The equivalence of items 3 and 4 gives a condition for the existence of κf in terms of μf and νf and reduces calculation of κf to that of μf and νf. This is already helpful as μf and νf are based on the implication order and therefore often easier to obtain than κf directly. In the following we aim to establish an even closer connection, namely $\kappa f = \nu f$, under suitable conditions.

3   Overall Strategy

In the remainder of this paper we further study how the $\sqsubseteq$-least fixpoint κf and the $\sqsubseteq$-greatest fixpoint νf of a function f are related when f represents computations with progress. We will give sufficient conditions for $\kappa f = \nu f$. Corollary 11 will establish this equality by using Proposition 1. To this end, we will derive conditions under which $\nu f \sqsubseteq \mu f \sqcup \mathsf{L}$.

To establish $\nu f \sqsubseteq \mu f \sqcup \mathsf{L}$, the main idea is to bound the excess of νf over μf by a meet of the form $\bigsqcap a_i$ for suitable elements a_i. That is, we prove $\nu f \sqsubseteq \mu f \sqcup \bigsqcap a_i$ and $\bigsqcap a_i \sqsubseteq \mathsf{L}$. Corollary 10 will show the latter. The former is established by the general Theorem 3 and a series of specialisations in Corollaries 4, 5 and 7. The specialisations instantiate $a_i = b^i \top$ for a suitable element b.

While our unified computation model supports a notion of progress, we still need a way to study different kinds of progress and to ensure a given computation actually makes progress. The idea for this is to separate the two concerns of progress and computation, which is based on having a separate time variable as described in [25–27]. To achieve this separation algebraically, we split the function f representing a recursion into a composition $f = g \circ h$. Here, the function h represents the computation done in the body of a recursion (without further recursive invocations) and the function g represents the progress made by this computation.

By imposing different conditions on the function g we can ensure actual progress and study different kinds of progress; this is shown in Theorem 9 and subsequently discussed in Sect. 5. Theorem 6 shows that the function h can represent computations carried out by programs constructed from choice, sequence, iteration and similar constructs. Specifically, we will consider the function $f(x) = a \cdot h(x)$, where progress is modelled by a suitable element a that satisfies $a \sqsubseteq b$ and is sequentially composed with the computation $h(x)$.

The development is split into two sections. We first cover results that can be derived algebraically in Sect. 4. In Sect. 5 we look closer into the computation model to discuss progress.

4   Connecting Fixpoints Algebraically

In this section we derive general results relating the $\sqsubseteq$-least and $\sqsubseteq$-greatest fixpoints of a function in the algebraic setting of Sect. 2.1.

We start with a simple upper bound for $\sqsubseteq$-greatest fixpoints. Kleene's recursion theorem gives the representation $\nu f = \bigsqcap_{i \in \mathbb{N}} f^i(\top)$ if the function f satisfies certain continuity requirements [9, 30, 32]. It is evident from previous proofs that one of the two inequalities comprising this equation only needs that f is $\sqsubseteq$-isotone.

Lemma 2. *Let S be a complete lattice. Let $f : S \to S$ be $\sqsubseteq$-isotone. Then $\nu f \sqsubseteq \bigsqcap_{i \in \mathbb{N}} f^i(\top)$.*

Proof. The claim follows if $\nu f \sqsubseteq f^i(\top)$ for each $i \in \mathbb{N}$. We show this by induction over i. The base case $i = 0$ follows since $\nu f \sqsubseteq \top$. The inductive case follows since for $i \in \mathbb{N}$ we have

$$\nu f = f(\nu f) \sqsubseteq f(f^i(\top)) = f^{i+1}(\top)$$

using that f is $\sqsubseteq$-isotone. $\square$

Next comes a general result that relates νf and μf by capturing the excess of $f^i(\top)$ over μf in the element a_i. It also splits the function f into a composition $g \circ h$ where an application of h does not increase the excess and an application of g may increase the excess from a_i to a_{i+1}. When we apply this result later, the component g will represent the progress of the computation.

Theorem 3. *Let S be a MID-complete lattice. Let $g : S \to S$ distribute over $\sqcup$ and let $h : S \to S$ be $\sqsubseteq$-isotone. Let $a_i \in S$ such that $g(a_i) \sqsubseteq a_{i+1}$ for each $i \in \mathbb{N}$. Assume $h(\top) \sqsubseteq a_0$ and $h(a_i \sqcup x) \sqsubseteq a_i \sqcup h(x)$ for each $i \geq 1$ and $x \in S$. Consider $f = g \circ h$. Then $\nu f \sqsubseteq \mu f \sqcup \bigsqcap_{i \geq 1} a_i$.*

Proof. We show by induction over i that $f^i(\top) \sqsubseteq a_i \sqcup \mu f$ for each $i \geq 1$. The base case $i = 1$ follows by

$$f(\top) = g(h(\top)) \sqsubseteq g(a_0) \sqsubseteq a_1 \sqsubseteq a_1 \sqcup \mu f$$

using that g is $\sqsubseteq$-isotone since it distributes over $\sqcup$. The inductive case follows since for $i \geq 1$ we have

$$f^{i+1}(\top) = f(f^i(\top)) = g(h(f^i(\top))) \sqsubseteq g(h(a_i \sqcup \mu f)) \sqsubseteq g(a_i \sqcup h(\mu f))$$
$$= g(a_i) \sqcup g(h(\mu f)) \sqsubseteq a_{i+1} \sqcup f(\mu f) = a_{i+1} \sqcup \mu f$$

using the assumption $h(a_i \sqcup x) \sqsubseteq a_i \sqcup h(x)$ with $x = \mu f$. The overall claim then follows by

$$\nu f \sqsubseteq \bigsqcap_{i \in \mathbb{N}} f^i(\top) \sqsubseteq \bigsqcap_{i \geq 1} f^i(\top) \sqsubseteq \bigsqcap_{i \geq 1}(a_i \sqcup \mu f) = \mu f \sqcup \bigsqcap_{i \geq 1} a_i$$

using Lemma 2. □

The following corollaries specialise the previous result. We first instantiate the excess to $a_i = b^i c$ for suitable elements b and c, and the function g to sequential composition of an element a such that $a \sqsubseteq b$. The latter condition guarantees that an application of g increases the excess from a_i to at most a_{i+1}.

The intuition is that b captures progress made by the body of the recursion (without further invocations). Hence $b^i c$ represents progress up to the ith unfolding of the recursion where c contains potential progress by further unfoldings.

Like b, the element a captures progress of the computation. Distinguishing a and b in the following results makes it possible later to establish $h(b^i c \sqcup x) \sqsubseteq b^i c \sqcup h(x)$, which separates the progress part $b^i c$ from the remainder of the computation $h(x)$. Setting $a = b$ would work in many models, but in trace models the element a represents specific progress that does not commute with other parts of the computation, so cannot be separated this way. However, we are able to choose an upper bound b on the progress that does commute and therefore can be collected separately. This will be demonstrated in more detail in Sect. 5.

Corollary 4. *Let S be a MID-complete lattice-ordered semiring. Let $h : S \to S$ be $\sqsubseteq$-isotone. Let $a, b, c \in S$ such that $a \sqsubseteq b$ and $h(\top) \sqsubseteq c$ and $h(b^i c \sqcup x) \sqsubseteq b^i c \sqcup h(x)$ for each $i \geq 1$ and $x \in S$. Consider $f(x) = a \cdot h(x)$. Then $\nu f \sqsubseteq \mu f \sqcup \bigsqcap_{i \geq 1} b^i c$.*

Proof. Let $g(x) = ax$ and $a_i = b^i c$ for each $i \in \mathbb{N}$. Then g distributes over $\sqcup$ and $h(\top) \sqsubseteq c = a_0$ and

$$g(a_i) = g(b^i c) = ab^i c \sqsubseteq bb^i c = b^{i+1} c = a_{i+1}$$

for each $i \in \mathbb{N}$. Moreover $h(a_i \sqcup x) \sqsubseteq a_i \sqcup h(x)$ for each $i \geq 1$ and $x \in S$. Finally $f = g \circ h$. Thus the claim follows by Theorem 3. □

We next instantiate $c = \top$, which simplifies the statement of the previous result and is sufficient for the development in this paper. The intuition for setting $c = \top$ is to make no assumptions in the ith unfolding of the recursion about progress that may happen in further recursive invocations.

Corollary 5. *Let S be a MID-complete lattice-ordered semiring. Let $h : S \to S$ be $\sqsubseteq$-isotone. Let $a, b \in S$ such that $a \sqsubseteq b$ and $h(b^i \top \sqcup x) \sqsubseteq b^i \top \sqcup h(x)$ for each $i \geq 1$ and $x \in S$. Consider $f(x) = a \cdot h(x)$. Then $\nu f \sqsubseteq \mu f \sqcup \bigsqcap_{i \geq 1} b^i \top$.*

Proof. Let $c = \top$. Then $h(\top) \sqsubseteq c$ and $h(b^i c \sqcup x) \sqsubseteq b^i c \sqcup h(x)$ for each $i \geq 1$ and $x \in S$. Thus the claim follows by Corollary 4. $\square$

The elements $b^i \top$ for $i \geq 1$ form a chain since $b^{i+1} \top = b^i b \top \sqsubseteq b^i \top$.

We still need to ensure that an application of h does not increase the excess, that is, $h(a_i \sqcup x) \sqsubseteq a_i \sqcup h(x)$ in Theorem 3. The next result gives sufficient conditions for an element a_i to satisfy this property.

The ideas underlying the conditions $a_i x \sqsubseteq a_i$ and $x a_i \sqsubseteq a_i \sqcup x \bot$ are as follows. The element $a_i = b^i \top$ represents the progress b^i made by the first i unfoldings of the recursion followed by arbitrary behaviour $\top$ in further unfoldings. The composition $a_i x$ makes another computation x at the end, but this does not add anything new to the arbitrary behaviour $\top$ already available at the end of the ith unfolding, which results in $a_i x \sqsubseteq a_i$.

In contrast, the composition $x a_i$ makes another computation x at the start, which can have an overall effect. Observe that among the executions contained in x, only the finite ones will reach the computation a_i. The infinite, aborting and incomplete executions of x absorb any subsequent computation and are contained in $x \bot$. The element b is chosen to capture progress in a general way so that the finite executions of x commute with it. Hence this part of x can be postponed all the way until after the ith unfolding, where again it adds nothing new to the arbitrary behaviour $\top$. All executions together are therefore contained as per $x a_i \sqsubseteq a_i \sqcup x \bot$.

Theorem 6. *Let S be a lattice-ordered itering. Let $h : S \to S$ such that $h(x)$ is constructed from the parameter x, arbitrary constants and the operations $\sqcup$, $\sqcap$, $\cdot$ and $^\circ$. Let $c \in S$ such that $cx \sqsubseteq c$ and $xc \sqsubseteq c \sqcup x \bot$ for each $x \in S$. Then $h(c \sqcup x) \sqsubseteq c \sqcup h(x)$ for each $x \in S$.*

Proof. We prove the claim by induction over the structure of the expression defining h. The base case $h(x) = x$ follows by

$$h(c \sqcup x) = c \sqcup x = c \sqcup h(x)$$

The base case $h(x) = d$ for a constant $d \in S$ follows by

$$h(c \sqcup x) = d \sqsubseteq c \sqcup d = c \sqcup h(x)$$

The inductive case $h(x) = f(x) \sqcup g(x)$ follows by

$$h(c \sqcup x) = f(c \sqcup x) \sqcup g(c \sqcup x) \sqsubseteq c \sqcup f(x) \sqcup c \sqcup g(x) = c \sqcup f(x) \sqcup g(x) = c \sqcup h(x)$$

The inductive case $h(x) = f(x) \sqcap g(x)$ follows by

$$h(c \sqcup x) = f(c \sqcup x) \sqcap g(c \sqcup x) \sqsubseteq (c \sqcup f(x)) \sqcap (c \sqcup g(x)) = c \sqcup (f(x) \sqcap g(x)) = c \sqcup h(x)$$

The inductive case $h(x) = f(x)g(x)$ follows by

$$h(c \sqcup x) = f(c \sqcup x)g(c \sqcup x) \sqsubseteq (c \sqcup f(x))(c \sqcup g(x))$$
$$= c(c \sqcup g(x)) \sqcup f(x)c \sqcup f(x)g(x)$$
$$\sqsubseteq c \sqcup c \sqcup f(x)\bot \sqcup f(x)g(x) = c \sqcup f(x)g(x) = c \sqcup h(x)$$

using the assumption $cy \sqsubseteq c$ with $y = c \sqcup g(x)$ and the assumption $zc \sqsubseteq c \sqcup z\bot$ with $z = f(x)$.

The inductive case $h(x) = f(x)^\circ$ follows by

$$h(c \sqcup x) = f(c \sqcup x)^\circ \sqsubseteq (c \sqcup f(x))^\circ = f(x)^\circ (c \cdot f(x)^\circ)^\circ \sqsubseteq f(x)^\circ c^\circ$$
$$= f(x)^\circ (1 \sqcup cc^\circ) \sqsubseteq f(x)^\circ (1 \sqcup c) = f(x)^\circ \sqcup f(x)^\circ c$$
$$\sqsubseteq f(x)^\circ \sqcup c \sqcup f(x)^\circ \bot = c \sqcup f(x)^\circ = c \sqcup h(x)$$

using sumstar and unfold properties of $^\circ$, the assumption $cy \sqsubseteq c$ with $y = f(x)^\circ$ and with $y = c^\circ$ and the assumption $zc \sqsubseteq c \sqcup z\bot$ with $z = f(x)^\circ$.   □

This means that the results of this section apply to any recursion whose characteristic function h is composed of constants, $\sqcup$, $\sqcap$, $\cdot$ and $^\circ$. In particular, the body of the recursion can use sequential compositions, conditionals and while-loops, the semantics of which are based on $\sqcup$, $\cdot$ and $^\circ$.

The following result combines Corollary 5 and Theorem 6. It provides the interface to the model-based discussion in Sect. 5.

Corollary 7. *Let S be a MID-complete lattice-ordered itering. Let $h : S \to S$ such that $h(x)$ is constructed from the parameter x, arbitrary constants and the operations $\sqcup$, $\sqcap$, $\cdot$ and $^\circ$. Let $a, b \in S$ such that $a \sqsubseteq b$ and $xb^i\top \sqsubseteq b^i\top \sqcup x\bot$ for each $i \geq 1$ and $x \in S$. Consider $f(x) = a \cdot h(x)$. Then $\nu f \sqsubseteq \mu f \sqcup \sqcap_{i \geq 1} b^i\top$.*

Proof. We apply Theorem 6 with $c = b^i\top$ for each $i \geq 1$. For this, first observe that $cx = b^i\top x \sqsubseteq b^i\top = c$ for each $x \in S$ since $\top x \sqsubseteq \top$ and composition is $\sqsubseteq$-isotone. Second, $xc \sqsubseteq c \sqcup x\bot$ for each $x \in S$ by the assumption of the present corollary. Hence $h(c \sqcup x) \sqsubseteq c \sqcup h(x)$ for each $x \in S$ by Theorem 6. Moreover, the function h is $\sqsubseteq$-isotone since it is composed of $\sqsubseteq$-isotone constructs. Thus the claim follows by Corollary 5.   □

We conclude this section with the following version of Theorem 3, which applies to functions of the form $f' = h \circ g$ instead of $f = g \circ h$. Other than this swap, the assumptions and conclusions of the two theorems are the same. The proof is by rolling fixpoints.

Theorem 8. *Let S be a MID-complete lattice. Let $g : S \to S$ distribute over $\sqcup$ and let $h : S \to S$ be $\sqsubseteq$-isotone. Let $a_i \in S$ such that $g(a_i) \sqsubseteq a_{i+1}$ for each $i \in \mathbb{N}$. Assume $h(\top) \sqsubseteq a_0$ and $h(a_i \sqcup x) \sqsubseteq a_i \sqcup h(x)$ for each $i \geq 1$ and $x \in S$. Consider $f' = h \circ g$. Then $\nu f' \sqsubseteq \mu f' \sqcup \sqcap_{i \geq 1} a_i$.*

Proof. Let $f = g \circ h$. Then $\nu f' = \nu(h \circ g) = h(\nu(g \circ h)) = h(\nu f)$ by the rolling property of ν. Similarly $\mu f' = h(\mu f)$ by the rolling property of μ. Moreover for each $i \geq 1$ we obtain

$$h(\mu f \sqcup \textstyle\bigsqcap_{j \geq 1} a_j) \sqsubseteq h(\mu f \sqcup a_i) \sqsubseteq a_i \sqcup h(\mu f)$$

using the assumption $h(a_i \sqcup x) \sqsubseteq a_i \sqcup h(x)$ with $x = \mu f$. Hence

$$\nu f' = h(\nu f) \sqsubseteq h(\mu f \sqcup \textstyle\bigsqcap_{j \geq 1} a_j) \sqsubseteq \textstyle\bigsqcap_{i \geq 1}(a_i \sqcup h(\mu f)) = h(\mu f) \sqcup \textstyle\bigsqcap_{i \geq 1} a_i$$
$$= \mu f' \sqcup \textstyle\bigsqcap_{i \geq 1} a_i$$

using Theorem 3. □

5 Connecting Fixpoints of Computations with Progress

In this section we work in the more detailed setting of the computation model presented in Sect. 2.2. This allows us to discuss progress more precisely.

In order to apply Corollary 7 we need to choose b such that $x b^i \top \sqsubseteq b^i \top \sqcup x \bot$ for each $x \in S$ and $i \geq 1$. As the following result shows, this condition holds for each computation $b = (\mathsf{O}|\mathsf{O}|\mathsf{O}|B)$ such that $\mathsf{F}B \subseteq B\mathsf{F}$.

Because we have separated the progress part from the actual computation in the recursion, we can now focus solely on the progress part. The intuition is that b is a general computation which captures a finite amount of progress. In particular, b does not contain any incomplete, aborting or infinite executions, which gives the form $(\mathsf{O}|\mathsf{O}|\mathsf{O}|B)$ for a relation B. Typically B will affect only the part of the state representing progress (such as a clock or a trace). The relation F describes an arbitrary amount of finite progress and allows arbitrary changes to the remaining part of the state. In most instances discussed below, B and F actually commute, but this is not necessary for the following result.

Theorem 9. *Let S be the computations in our model. Let $b = (\mathsf{O}|\mathsf{O}|\mathsf{O}|B)$ for a relation $B \subseteq \mathsf{F}$ with $\mathsf{F}B \subseteq B\mathsf{F}$. Then $x b^i \top \sqsubseteq b^i \top \sqcup x \bot$ for each $x \in S$ and $i \geq 1$.*

Proof. We first prove $b^i = (\mathsf{O}|\mathsf{O}|\mathsf{O}|B^i)$ for each $i \geq 1$ by induction over i. The base case $i = 1$ follows immediately. The inductive case $i \geq 1$ holds by

$$b^{i+1} = b b^i = (\mathsf{O}|\mathsf{O}|\mathsf{O}|B)(\mathsf{O}|\mathsf{O}|\mathsf{O}|B^i) = (\mathsf{O}|\mathsf{O}|\mathsf{O}|BB^i) = (\mathsf{O}|\mathsf{O}|\mathsf{O}|B^{i+1})$$

Hence for each $i \geq 1$ we have

$$b^i \top = (\mathsf{O}|\mathsf{O}|\mathsf{O}|B^i)(\mathsf{F}|\mathsf{F}|\mathsf{F}_\infty|\mathsf{F}) = (B^i\mathsf{F}|B^i\mathsf{F}|B^i\mathsf{F}_\infty|B^i\mathsf{F})$$

We next prove that $\mathsf{F}B \subseteq B\mathsf{F}$ implies $\mathsf{F}B^i \subseteq B^i\mathsf{F}$ for each $i \geq 0$ by induction over i. The base case $i = 0$ follows by

$$\mathsf{F}B^0 = \mathsf{F}\mathsf{I} = \mathsf{F} = \mathsf{I}\mathsf{F} = B^0\mathsf{F}$$

The inductive case $i \geq 0$ follows by

$$FB^{i+1} = FB^i B \subseteq B^i FB \subseteq B^i BF = B^{i+1}F$$

Thus for arbitrary $x = (N|P|Q|R) \in S$ and $i \geq 1$ we obtain

$$
\begin{aligned}
xb^i\top &= (N|P|Q|R)(B^i\mathsf{F}|B^i\mathsf{F}|B^i\mathsf{F}_\infty|B^i\mathsf{F}) \\
&\sqsubseteq (N|P|Q|\mathsf{F})(B^i\mathsf{F}|B^i\mathsf{F}|B^i\mathsf{F}_\infty|B^i\mathsf{F}) \\
&= (N \cup FB^i\mathsf{F}|P \cup FB^i\mathsf{F}|Q \cup FB^i\mathsf{F}_\infty|FB^i\mathsf{F}) \\
&\sqsubseteq (N \cup B^i\mathsf{FF}|P \cup B^i\mathsf{FF}|Q \cup B^i\mathsf{FF}_\infty|B^i\mathsf{FF}) \\
&= (N \cup B^i\mathsf{F}|P \cup B^i\mathsf{F}|Q \cup B^i\mathsf{F}_\infty|B^i\mathsf{F}) \\
&= (B^i\mathsf{F}|B^i\mathsf{F}|B^i\mathsf{F}_\infty|B^i\mathsf{F}) \cup (N|P|Q|\mathsf{O}) \\
&= (B^i\mathsf{F}|B^i\mathsf{F}|B^i\mathsf{F}_\infty|B^i\mathsf{F}) \cup (N \cup RO|P \cup RO|Q \cup RO|RO) \\
&= (B^i\mathsf{F}|B^i\mathsf{F}|B^i\mathsf{F}_\infty|B^i\mathsf{F}) \cup (N|P|Q|R)(\mathsf{O}|\mathsf{O}|\mathsf{O}|\mathsf{O}) \\
&= b^i\top \sqcup x\bot
\end{aligned}
$$

using the progress requirement $R \subseteq \mathsf{F}$ and $\mathsf{F}B^i \subseteq B^i\mathsf{F}$.   □

We next consider if the expression $\sqcap_{i \geq 1} b^i\top$, which appears in Corollary 7, can be bounded by L. The following result gives a sufficient criterion based on the intersection $B' = \bigcap_{i \geq 1} B^i\mathsf{F}$.

Corollary 10. *Let S be the computations in our model. Let $h : S \to S$ such that $h(x)$ is constructed from the parameter x, arbitrary constants and the operations $\sqcup$, $\sqcap$, $\cdot$ and $^\circ$. Let $a \sqsubseteq b = (\mathsf{O}|\mathsf{O}|\mathsf{O}|B)$ for a relation $B \subseteq \mathsf{F}$ with $FB \subseteq BF$ and $\bigcap_{i \geq 1} B^i\mathsf{F} = \mathsf{O}$. Consider $f(x) = a \cdot h(x)$. Then $\nu f \sqsubseteq \mu f \sqcup \mathsf{L}$.*

Proof. By Corollary 7 and Theorem 9 we obtain $\nu f \sqsubseteq \mu f \sqcup \sqcap_{i \geq 1} b^i\top$. From this the claim follows by

$$
\begin{aligned}
\sqcap_{i \geq 1} b^i\top &= \sqcap_{i \geq 1}(B^i\mathsf{F}|B^i\mathsf{F}|B^i\mathsf{F}_\infty|B^i\mathsf{F}) \sqsubseteq \sqcap_{i \geq 1}(B^i\mathsf{F}|B^i\mathsf{F}|\mathsf{F}_\infty|B^i\mathsf{F}) \\
&= (\textstyle\bigcap_{i \geq 1} B^i\mathsf{F}|\bigcap_{i \geq 1} B^i\mathsf{F}|\bigcap_{i \geq 1}\mathsf{F}_\infty|\bigcap_{i \geq 1} B^i\mathsf{F}) = (\mathsf{O}|\mathsf{O}|\mathsf{F}_\infty|\mathsf{O}) \sqsubseteq \mathsf{L}
\end{aligned}
$$

using that $B^i\mathsf{F}_\infty \subseteq \mathsf{F}_\infty$ by the progress requirement on that component.   □

The value of B' depends on the kind of progress represented by the computation model. We discuss several examples. We first consider examples which satisfy $a = b$ and then examples where $a = (\mathsf{O}|\mathsf{O}|\mathsf{O}|A)$ for $A \subset B$.

In the Boolean-time model, $\mathsf{F} = \top$ holds, so the assumption $FB \subseteq BF$ is equivalent to $\top B \subseteq B\top$, which implies $\top B\top \subseteq B\top\top = B\top$. By the Tarski property of relations [35], either $B = \mathsf{O}$ or $\top B\top = \top$, in which case $B\top = \top$ follows. If $B = \mathsf{O}$, we have $a = b = \bot$ and therefore $f(x) = \bot$, whence Corollary 10 is vacuous. If $B\top = \top$, we have $B^i\top = \top$ for each $i \geq 1$ by induction, so $B' = \bigcap_{i \geq 1} B^i\top = \bigcap_{i \geq 1} \top = \top \neq \mathsf{O}$, whence Corollary 10 does not apply. The notion of progress in the Boolean-time model is too coarse for this result.

The abstract-time model counts steps and here $F = \{((v,t),(v',t')) \mid t \leq t'\}$ specifies that the counter, which is a natural number, does not decrease. To achieve strict progress, we consider

$$B = \{((v,t),(v,t')) \mid t' - t > 0\} = \{((v,t),(v,t')) \mid t' - t \geq 1\}$$

specifying that the state remains unchanged but the counter must increase. It follows that

$$FB = BF = \{((v,t),(v',t')) \mid t' - t \geq 1\}$$

specifying that the counter must increase. Moreover,

$$B^i = \{((v,t),(v,t')) \mid t' - t \geq i\}$$

for each $i \geq 1$ by induction, specifying that the state remains unchanged and the counter increases by at least i. Hence

$$B^i F = \{((v,t),(v',t')) \mid t' - t \geq i\}$$

for each $i \geq 1$, specifying that the counter increases by at least i. Therefore $B' = \bigcap_{i \geq 1} B^i F = O$ since any finite difference $t' - t$ will be exceeded after sufficiently many steps. Thus Corollary 10 applies.

In the real-time model we again have $F = \{((v,t),(v',t')) \mid t \leq t'\}$, but the previous argument fails. Specifically, strict progress using

$$B = \{((v,t),(v,t')) \mid t' - t > 0\}$$

implies that

$$FB = BF = \{((v,t),(v',t')) \mid t' - t > 0\}$$

and $B^2 = B$, whence $B^i = B$ for each $i \geq 1$ by induction. Hence $B^i F = BF$ for each $i \geq 1$. It follows that $B' = BF \neq O$, so Corollary 10 does not apply. More precisely, we obtain only $\bigcap_{i \geq 1} b^i \top \sqsubseteq (BF|BF|F_\infty|BF)$, which is not below L.

However, if strict progress means at least c units of time pass for an arbitrary $c > 0$, that is,

$$B = \{((v,t),(v,t')) \mid t' - t \geq c\}$$

we obtain

$$B^i = \{((v,t),(v,t')) \mid t' - t \geq ic\}$$

for each $i \geq 1$ by induction. Moreover,

$$FB = BF = \{((v,t),(v',t')) \mid t' - t \geq c\}$$

and

$$B^i F = \{((v,t),(v',t')) \mid t' - t \geq ic\}$$

reflecting that at least ic units of time pass after i steps. In this case, we obtain $B' = \bigcap_{i \geq 1} B^i F = O$ again, whence Corollary 10 applies.

So, in the real-time model the argument fails because there is no lower bound on the progress in each step. A particular instance of this is the Zeno effect where each step takes half the time of the preceding step. The argument works if we assume a uniform lower bound for all steps.

We remark that it is also possible to specify an upper bound on the amount of progress. One way to see this is to consider for an arbitrary $d \geq c$ the relation

$$A = \{((v,t),(v,t')) \mid d \geq t' - t \geq c\}$$

whence $A \subset B$, and use the previous argument for B. This includes the case $c = d$ where the amount of progress in each step is exactly specified.

In the trace model, $F = \preceq$ is the prefix relation on traces. We can achieve strict progress, for example, by

$$A = \{(tr, tr') \mid tr' = tr + [last(tr)]\}$$

which appends to the sequence tr its last element. Hence $A \subset B$ for the strict prefix relation $B = \prec$, which specifies that the trace gets longer by at least one element. It follows that $FB = B = BF$. Moreover B^i specifies that the trace gets longer by at least i elements. It follows that $B^iF = B^i$. Thus $B' = \bigcap_{i \geq 1} B^i = O$ since any finite length will be exceeded after sufficiently many steps. Therefore Corollary 10 applies.

The previous example also demonstrates why Corollaries 4, 5, 7 and 10 distinguish between a and b. Namely, to apply these results if $a = b$ we would need the condition $FA \subseteq AF$, which does not hold in this case. This is because F may extend the trace in an arbitrary way while A appends a specific element. However, A refines the more general B, which commutes with F.

We finally use Corollary 10 to connect least fixpoints in the approximation order with greatest fixpoints in the implication order.

Corollary 11. *Let S be the computations in our model. Let $h : S \to S$ such that $h(x)$ is constructed from the parameter x, arbitrary constants and the operations $\sqcup, \cdot$ and $\circ$. Let $a \sqsubseteq b = (O|O|O|B)$ for a relation $B \subseteq F$ with $FB \subseteq BF$ and $\bigcap_{i \geq 1} B^iF = O$. Consider $f(x) = a \cdot h(x)$. Then κf exists and $\kappa f = \nu f$.*

Proof. By item 2 of Proposition 1, the functions f and h are $\sqsubseteq$-isotone. We show the condition in item 4 of Proposition 1. By Corollary 10, we have $\nu f \sqsubseteq \mu f \sqcup L$. Hence

$$c(\nu f) = n(L)\top \sqcap \nu f \sqsubseteq \nu f = \nu f \sqcap (\mu f \sqcup L) = (\nu f \sqcap \mu f) \sqcup (\nu f \sqcap L)$$
$$= (\nu f \sqcap L) \sqcup \mu f \sqsubseteq (\nu f \sqcap L) \sqcup \mu f \sqcup n(\nu f)\top$$

By item 3 of Proposition 1, it follows that κf exists and $\kappa f = (\nu f \sqcap L) \sqcup \mu f$. Hence $\kappa f = \nu f$ using the previous calculation.   □

The assumptions of Corollary 11 are the same as those of Corollary 10 except the operation $\sqcap$ cannot be used in the construction of h, since it is not $\sqsubseteq$-isotone. The discussion following Corollary 10 applies in the same way.

6   Conclusion

In this paper we have studied when least fixpoints in the approximation order coincide with greatest fixpoints in the implication order. To achieve this, we separately captured the progress part of a recursive computation, an idea rooted in the time variables of [25–27]. The separation was achieved using algebraic means in a setting based on lattices and semirings, which unifies many computation models. We were then able to focus on the elements representing progress in a suitable unified computation model. This allowed us to state conditions under which the studied fixpoints coincide and to discuss the meaning of these conditions for various kinds of progress.

Two anonymous referees asked about necessary conditions for $\kappa f = \nu f$. Since Corollary 11 makes several assumptions, this question can be considered in many different ways. We briefly discuss a negative answer leaving further investigation to future work.

Section 5 shows that the assumption $\bigcap_{i \geq 1} B^i \mathsf{F} = \mathsf{O}$ of Corollary 11 fails if there is no lower bound for the progress B made by the body of the recursion in the real-time model. But even then, if the body of the recursion $f(x)$ makes arbitrary finite progress followed by an error (without a recursive invocation of x), the characteristic function f is constant and therefore has a unique fixpoint. Hence κf exists and $\kappa f = \nu f$ in this case.

We finally mention connections to a number of related works, which can also be explored further.

The semantics of iteration in a timed computation model (an extension of UTP designs by a time variable) is defined in [24] by inserting a statement that increments the clock before the tail-recursive call. This can be seen as an instance of separating the progress part of a computation. Inserting before the recursive call corresponds to the pattern of Theorem 8.

Corollary 11 relies on the property $\kappa f = (\nu f \sqcap \mathsf{L}) \sqcup \mu f$ of the $\sqsubseteq$-least fixpoint of f. This property holds not only in the computation model discussed in this paper, but in a number of other relational, matrix-based and multirelational computation models [19, 20].

The implication and approximation orders are related to the truth and knowledge/information orders used in the bilattice approach to the semantics of logic programs [13, 15].

Acknowledgement. I thank the anonymous referees for their helpful comments.

References

1. Aarts, C.J., et al.: Fixed-point calculus. Inf. Process. Lett. **53**(3), 131–136 (1995)
2. Alexandru, A., Ciobanu, G.: Abstract interpretations in the framework of invariant sets. Fundamenta Informaticae **144**(1), 1–22 (2016)
3. Back, R.J.R., von Wright, J.: Reasoning algebraically about loops. Acta Inf. **36**(4), 295–334 (1999)

4. de Bakker, J.W.: Semantics and termination of nondeterministic recursive programs. In: Michaelson, S., Milner, R. (eds.) ICALP 1976, pp. 435–477. Edinburgh University Press (1976)
5. Broy, M., Gnatz, R., Wirsing, M.: Semantics of nondeterministic and noncontinuous constructs. In: Bauer, F.L., Broy, M. (eds.) Program Construction. LNM, vol. 69, pp. 553–592. Springer, Heidelberg (1979). https://doi.org/10.1007/BFb0014683
6. Cohen, E.: Separation and reduction. In: Backhouse, R., Oliveira, J.N. (eds.) MPC 2000. LNCS, vol. 1837, pp. 45–59. Springer, Heidelberg (2000). https://doi.org/10.1007/10722010_4
7. Conway, J.H.: Regular Algebra and Finite Machines. Chapman and Hall, London (1971)
8. Cousot, P., Cousot, R.: An abstract interpretation framework for termination. In: POPL 2012, pp. 245–257. ACM (2012)
9. Davey, B.A., Priestley, H.A.: Introduction to Lattices and Order, 2nd edn. Cambridge University Press, Cambridge (2002)
10. Dunne, S.: Recasting Hoare and He's Unifying Theory of Programs in the context of general correctness. In: Butterfield, A., Strong, G., Pahl, C. (eds.) 5th Irish Workshop on Formal Methods. Electronic Workshops in Computing. The British Computer Society (2001)
11. Dunne, S.: Conscriptions: a new relational model for sequential computations. In: Wolff, B., Gaudel, M.-C., Feliachi, A. (eds.) UTP 2012. LNCS, vol. 7681, pp. 144–163. Springer, Heidelberg (2013). https://doi.org/10.1007/978-3-642-35705-3_7
12. Egli, H.: A mathematical model for non-deterministic computations. Technical report, Forschungsinstitut für Mathematik ETH Zürich (1975)
13. Fitting, M.: Bilattices and the semantics of logic programming. J. Logic Program. 11(2), 91–116 (1991)
14. Freyd, P.J., Ščedrov, A.: Categories, Allegories, North-Holland Mathematical Library, vol. 39. Elsevier Science Publishers (1990)
15. Ginsberg, M.L.: Multivalued logics: a uniform approach to reasoning in artificial intelligence. Comput. Intell. 4(3), 265–316 (1988)
16. Gordon, M.J.C.: The Denotational Description of Programming Languages. Springer, New York (1979)
17. Guttmann, W.: Algebras for iteration and infinite computations. Acta Inf. 49(5), 343–359 (2012)
18. Guttmann, W.: Extended conscriptions algebraically. In: Höfner, P., Jipsen, P., Kahl, W., Müller, M.E. (eds.) RAMICS 2014. LNCS, vol. 8428, pp. 139–156. Springer, Cham (2014). https://doi.org/10.1007/978-3-319-06251-8_9
19. Guttmann, W.: Multirelations with infinite computations. J. Logical Algebraic Methods Program. 83(2), 194–211 (2014)
20. Guttmann, W.: Infinite executions of lazy and strict computations. J. Logical Algebraic Methods Program. 84(3), 326–340 (2015)
21. Guttmann, W.: An algebraic approach to computations with progress. J. Logical Algebraic Methods Program. 85(4), 520–539 (2016)
22. Guttmann, W., Möller, B.: Normal design algebra. J. Logic Algebraic Program. 79(2), 144–173 (2010)
23. Hayes, I.J., Dunne, S.E., Meinicke, L.: Unifying theories of programming that distinguish nontermination and abort. In: Bolduc, C., Desharnais, J., Ktari, B. (eds.) MPC 2010. LNCS, vol. 6120, pp. 178–194. Springer, Heidelberg (2010). https://doi.org/10.1007/978-3-642-13321-3_12
24. Hayes, I.J., Dunne, S.E., Meinicke, L.A.: Linking unifying theories of program refinement. Sci. Comput. Program. 78(11), 2086–2107 (2013)

25. Hehner, E.C.R.: Real-time programming. Inf. Process. Lett. **30**(1), 51–56 (1989)
26. Hehner, E.C.R.: Termination is timing. In: van de Snepscheut, J.L.A. (ed.) MPC 1989. LNCS, vol. 375, pp. 36–47. Springer, Heidelberg (1989). https://doi.org/10. 1007/3-540-51305-1_3
27. Hehner, E.C.R.: Abstractions of time. In: Roscoe, A.W. (ed.) A Classical Mind, chap. 12, pp. 191–210. Prentice Hall (1994)
28. Hoare, C.A.R., He, J.: Unifying Theories of Programming. Prentice Hall Europe (1998)
29. Höfner, P., Möller, B.: An algebra of hybrid systems. J. Logic Algebraic Program. **78**(2), 74–97 (2009)
30. Kleene, S.C.: Introduction to Metamathematics. North-Holland Publishing Company (1952)
31. Kozen, D.: A completeness theorem for Kleene algebras and the algebra of regular events. Inf. Comput. **110**(2), 366–390 (1994)
32. Manna, Z.: Mathematical Theory of Computation. McGraw-Hill, New York (1974)
33. Möller, B.: Kleene getting lazy. Sci. Comput. Program. **65**(2), 195–214 (2007)
34. Schmidt, G., Hattensperger, C., Winter, M.: Heterogeneous relation algebra. In: Brink, C., Kahl, W., Schmidt, G. (eds.) Relational Methods in Computer Science. ACS, pp. 39–53. Springer, Vienna (1997). https://doi.org/10.1007/978-3-7091-6510-2_3
35. Schmidt, G., Ströhlein, T.: Relationen und Graphen. Springer, Heidelberg (1989)
36. Shinwell, M.R.: The fresh approach: functional programming with names and binders. Ph.D. thesis, University of Cambridge (2005). Available as Technical report UCAM-CL-TR-618, Computer Laboratory, University of Cambridge
37. Stoy, J.E.: Denotational Semantics: The Scott-Strachey Approach to Programming Language Theory. MIT Press, Cambridge (1977)
38. Tarski, A.: A lattice-theoretical fixpoint theorem and its applications. Pac. J. Math. **5**(2), 285–309 (1955)
39. Vuillemin, J.: Correct and optimal implementations of recursion in a simple programming language. Technical report 24, IRIA (1973). Also in J. Comput. Syst. Sci. **9**(3), 332–354 (1974)
40. von Wright, J.: Towards a refinement algebra. Sci. Comput. Program. **51**(1–2), 23–45 (2004)

Probabilistic Semantics for RoboChart
A Weakest Completion Approach

Jim Woodcock[1]($\boxtimes$)(iD), Ana Cavalcanti[1](iD), Simon Foster[1](iD),
Alexandre Mota[2](iD), and Kangfeng Ye[1]

[1] University of York, York, UK
{jim.woodcock,ana.cavalcanti,simon.foster,kangfeng.ye}@york.ac.uk
[2] Federal University of Pernambuco, Recife, Brazil
acm@cin.ufpe.br

Abstract. We outline a probabilistic denotational semantics for the RoboChart language, a diagrammatic, domain-specific notation for describing robotic controllers with their hardware platforms and operating environments. We do this using a powerful (but perhaps not so well known) semantic technique: He, Morgan, and McIver's *weakest completion semantics*, which is based on Hoare and He's Unifying Theories of Programming. In this approach, we do the following: (1) start with the standard semantics for a nondeterministic programming language; (2) propose a new probabilistic semantic domain; (3) propose a forgetful function from the probabilistic semantic domain to the standard semantic domain; (4) use the converse of the forgetful function to embed the standard semantic domain in the probabilistic semantic domain; (5) demonstrate that this embedding preserves program structure; (6) define the probabilistic choice operator. Weakest completion semantics guides the semantic definition of new languages by building on existing semantics and, in this case, tackling a notoriously thorny issue: the relationship between demonic and probabilistic choice. Consistency ensures that programming intuitions, development techniques, and proof methods can be carried over from the standard language to the probabilistic one. We largely follow He et al., our contribution being an explication of the technique with meticulous proofs suitable for mechanisation in Isabelle/UTP.

Keywords: RoboChart language · Robotic controllers · Statecharts · Probabilistic semantics · Relational calculus · Unifying Theories of Programming (UTP) · Weakest completion semantics

1 Introduction

Modern robotics simulators enable fast prototyping of robots, using a virtual simulation environment as a software creation and design tool. They provide realistic, computer gaming-style, 3-D rendering of robots and environments with

© Springer Nature Switzerland AG 2019
P. Ribeiro and A. Sampaio (Eds.): UTP 2019, LNCS 11885, pp. 80–105, 2019.
https://doi.org/10.1007/978-3-030-31038-7_5

physics engines to animate their movements authentically in automatically generated movies. Examples of such simulators include the Virtual Robot Experimentation Platform (V-REP) [54] and Webots [56].

One drawback of using these simulators as part of a principled development process is a lack of tool interoperability, with each simulator depending on a customised programming language or API. As a result of this, there are few possibilities for reuse of specifications and algorithms, and software development starts at a low-level with few abstractions. A notable exception to this is the Robot Modeling Language, ROBOTML [7], which targets the design of robotic applications, their simulation, and their deployment to multiple target execution platforms. The motivation for ROBOTML is to encourage a more abstract design process with explicit architectures, but there is no support for formal methods for verifying properties of these designs and architectures.

The RoboStar programme is developing a framework for modelling and simulating mobile and autonomous robots [49].[1] An early product of the research is the RoboChart language, a graphical domain-specific notation with a code generator that automatically produces mathematical models [40,41,48] in the notations of Communicating Sequential Processes (CSP) [52]. This enables the analysis of structural properties of RoboCharts: freedom from deadlock, livelock, and nondeterminism; it also supports the verification of more general untimed and timed properties by refinement checking [51]. RoboChart has an associated Eclipse-based development support environment, RoboTool [39], that enables graphical modelling and automatic generation of CSP scripts, and is integrates CSP's refinement model checker FDR4 [9].

RoboCalc's RoboSim language [4] provides a second graphical notation for developing simulations. A novel feature of RoboSim is the ability to verify simulations against their abstract RoboChart models. This ensures that the combination of models, simulations, deployed controllers, and hardware platforms refine the properties verified and validated by analysis and simulation. RoboChart and RoboSim support real time, discrete, continuous, and probabilistic properties; we consider only discrete probabilistic semantics in this paper. Our probabilistic models are essentially Markov Decision Processes (MDPs).

RoboChart and RoboSim have strong mathematical foundations, but they are also practical for industrial-strength robotic software engineering. This requires that they be attractive to practising engineers, but with additional powers to enable formal verification. The state of the art in industry is to use modelling techniques to specify the behaviours of robot controllers, but not the robotic hardware platform or the operating environment. Even at their most advanced, current industrial techniques use only simple state machines without formal semantics [3,7,47,55]. Any abstract descriptions that are used guide simulation development, but without any relationship between abstract descriptions and implemented code. There is often a so-called "reality gap" between the state machine and simulation on the one hand, and the hardware platform on the other, and ad hoc adjustments must be made to get the robot working. It is for this reason that

---

[1] The RoboStar programme includes a number of individual projects, including Robo-Calc, which is developing a calculus of software engineering for robotic controllers.

we have developed RoboChart with high-fidelity modelling capabilities, including continuous time and probabilism [41].[2] There is little motivation to keep the abstract state machine in line with these changes.

RoboChart has a probabilistic choice operator, but this cannot be supported by the translation into CSP, because the standard semantics and tools are not probabilistic. This paper presents an approach to developing a suitable imperative, reactive, probabilistic semantics for RoboChart. The method chosen to develop this semantics is the weakest completion semantics [24] approach based on Unifying Theories of Programming [31]. In this paper, we consider a semantics for the imperative action language for RoboChart. Elsewhere, we consider the use of this semantics to produce a sound translation from diagrams to mathematics, suitable for analysis by verification tools [10].

Our contribution is an explication of the weakest completion approach: a detailed analysis of this principle for developing semantics, enabling future application to RoboChart [41], a complex language with events, timed primitives, rich data types, a concurrency model based on synchronous and asynchronous communications, and shared variables. The inspiration for our work is precisely that of He, Morgan, and McIver [24]; but it is not straightforward to take their informal proof outlines and use them directly in a mechanical theorem prover: they are inspirational, but essentially informal. We present an abbreviation of our proof due to space limitations, but our proof steps are based on explicit axioms, lemmas, theorems, and inferences.

This paper has the following structure. In Sect. 2, we describe a few elements of the RoboChart language. In Sect. 3, we give an overview of Unifying Theories of Programming. In Sect. 4, we provide an interlude, where we discuss two predicate transformers: weakest preconditions and weakest prespecifications. In Sect. 5, we describe the technique of weakest completion semantics. In Sect. 6, we present a nondeterministic probabilistic programming language and its semantic domain. In Sect. 7, we describe the semantics of probabilistic choice and discuss how to combine distributions. In Sect. 8, we provide a detailed example: embedding nondeterministic choice in the probabilistic domain. In Sect. 9, we discuss related work on formalising probabilistic RoboCharts. Finally, in Sect. 10, we draw some conclusions from this research in progress and discuss future work.

2   RoboChart

We model robot controllers using RoboChart [41], a UML profile [44]. RoboChart models are Statecharts, a diagrammatic notation for defining behaviour [22]. State machines are part of the fabric of computing, recognisable in many forms (including, for example, Mealy and Moore automata [37]) and they are widely accepted in the embedded-software industry as a design notation. Statecharts [22] extend these familiar diagrams with two features: hierarchy and

---

[2] The difficulty of transferring simulated experience into the real world, often called the "reality gap" [32], is a subtle but important discrepancy between reality and simulation that prevents simulated robotic experience from directly enabling effective real-world performance [2].

orthogonal regions. Hierarchy is provided by allowing states themselves to contain state machines, where control flow resides in exactly one state: the so-called OR-decomposition of behaviour; orthogonal regions provide a complementary AND-decomposition, where control-flow can simultaneously reside in one position in each orthogonal region, fully independently. This parallel decomposition can lead to a reduction in complexity (the "conjunction as composition" paradigm [63]). RoboChart inherits both features: it has hierarchical state machines (which encourage modularity and reuse) and parallelism, arising for components defined by several state machines and for composite states, including durative actions.

In RoboChart, the behaviour of a robot is characterised by a state, in which it may execute a particular operation and react to events from its environment. RoboChart includes structures for describing robotic platforms and their controllers, with CSP-style synchronous communication between controllers and asynchronous communication between controllers and their hardware. It has constructs to specify time properties: budgets and deadlines for operations and events. Here, we consider only the probabilistic aspects of RoboChart. We isolate a language subset of flat nondeterministic state machines with a probabilistic choice node.

RoboTool [39] provides a graphical editor for RoboChart models and automatically generates mathematical definitions in CSP that precisely define their behaviour. RoboTool is closely coupled with the FDR model checker [9] to analyse these definitions.

We present two RoboChart models to illustrate the language. The first model is part of the controller for a tele-operated robot used to search an arena for evidence of a harmful chemical, using the receptor density algorithm [28]. The RoboChart is depicted in Fig. 1, which is from a RoboTool session. The robot controller uses a sensor to detect changes in the chemical composition of air over time. It reacts to gas anomalies depending on their nature and composition: with a yellow or a red light, a siren, and a flag to mark the location. The hardware includes a robot body, wheels, and motor, a main processor to detect gas and accept movement commands from an operator, and a microcontroller to manage the light, siren, and flag.

Our second example is part of the controller for a foraging robot [57] (the corresponding RoboChart model and the results of its analysis are available online [50]).[3] The robot has an idealised randomising device with two states that are equally likely to occur; the device generates an outcome from a flip event in every time step.[4] The robot uses the device to decide whether to terminate or to

---

[3] The Statechart in this example is originally due to Jansen [34], but has been reinterpreted here as a robotics example.

[4] The semantics in this paper does not capture the real-time behaviour of RoboChart; however, every transition in an MDP takes unit time. When we develop the real-time probabilistic model, these two notions of time will be complementary, allowing events to be simultaneous with respect to the real-time clock, but ordered at the MDP level: super-dense time.

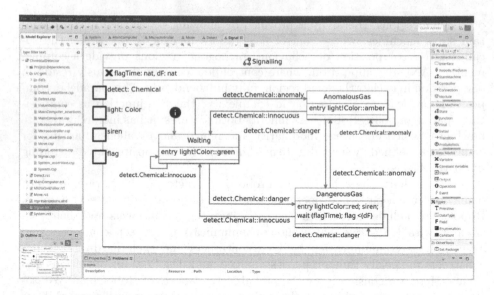

Fig. 1. Signalling state machine for Chemical Detector Robot (taken from [40]).

continue a particular activity (here, foraging for energy). The robot may choose to ignore the outcome of the device. Finally, the robot considers only a limited number of times whether to continue foraging. We call this number N and leave it loosely defined. Our simple modelling objective is to explore different values for N that give us a high probability of terminating.

We specify the behaviour of the device as a RoboChart model in Fig. 2. One possibility in the FORAGE state is for the flip event to occur and the robot to remain in the FORAGE state; this models the robot ignoring the randomising device. The other possibility is available only if the number of choices has not been exhausted ($flips < N$). In this case, the robot controller proceeds to a probabilistic choice between two equally likely alternatives. One alternative is to move into the STOP state, which it signals with the stop event; the other alternative is to return to the FORAGE state, signalling this with the forage event. In both cases, the controller keeps track of the number of choices made. Note that, if in the FORAGE state $flips < N$, then the behaviour is nondeterministic: the robot controller might take either alternative. In the STOP state, only the flip event is possible, with a self-loop acting as a sink. A well-formed MDP must be free from deadlock (every state must have at least one outgoing transition), and anyway, this transition is needed because of the requirement that flip must occur in every time step, even when the controller has terminated.

Analysis of the generated model using PRISM [35] shows that the model is deadlock free, but that the STOP state is not always possible (the minimum probability of finally reaching STOP is zero) because the model could stay in the FORAGE state forever. Additionally, using experiment for N ranging from 1

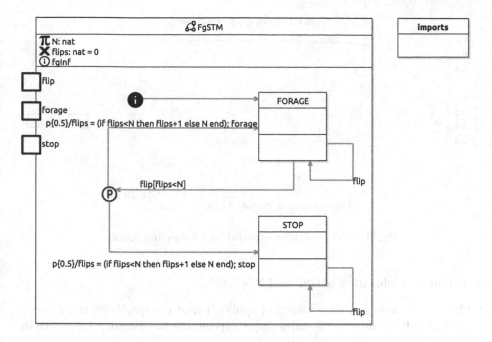

Fig. 2. RoboChart model of a foraging robot.

to 20, we can obtain the probability of finally reaching STOP, as shown in Fig. 3. For N ≥ 6 the device will terminate with probability greater than 0.98.

3   Unifying Theories of Programming

There are tutorial introductions to UTP's theory of designs [59,60], CSP [6], and the use of Galois connections to link these theories [61]. UTP embodies Hehner's predicative semantic paradigm [25–27], where programs are predicates [29]: a program is identified with its meaning as a predicate, expressed pointwise. Theories describe the meaning of a computation as a relation between a before-state and an after-state, and these relations form complete lattices ordered by refinement. Several basic UTP theories are relevant to this paper.

1. A relational theory of a nondeterministic programming language (basically, Dijkstra's guarded command language (GCL)) supports reasoning about partial correctness [31, Chap. 2].
2. A theory of designs, pre- and postcondition pairs, and an associated version of GCL supports reasoning about total correctness [31, Chap. 3].
3. A theory of reactive processes with communication and concurrency [31, Chap. 8].
4. A theory of CSP, essentially a predicative version of CSP's failures-divergences semantics [31, Chap. 8].

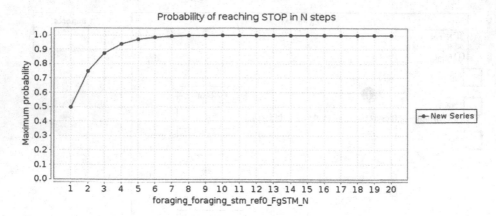

Fig. 3. Model checking experiment for foraging robot.

5. Circus, a combination of CSP and Z [45, 46].

UTP has been used in a wide variety of applications, from specifying and reasoning about difficult program features [23], to specifying the semantic interfaces in a cyber-physical systems tool chain [12, 36].

A core concept is the embedding of the pre- and postconditions of designs in other semantic domains. For example, the theory of reactive designs [6] is an embedding of designs in the theory of reactive processes, which brings the familiar techniques of assertional reasoning and design calculi to reactive programming, allowing the creation of a reactive Hoare logic and a reactive weakest precondition calculus.

Unification in UTP is in three dimensions:

1. Programming paradigms: comparing and combining different language features in a coherent way.
2. Levels of abstraction: refining different design concepts.
3. Methods of presentation: moving between denotational, algebraic, and operational semantics.

There are four principal mechanisms for unification:

1. Subset embeddings, e.g., total and partial correctness (designs and relations) [58].
2. Weakest completion semantics, e.g., probabilistic and standard programs, as explained in Sect. 5.
3. Galois connections, e.g., imperative programs and reactive processes [5, 58].
4. Parametrised theories, e.g., reactive processes and hybrid processes [15].

We have implemented UTP in the Isabelle/HOL theorem prover [42]. The resulting proof tool is Isabelle/UTP [13, 16–19]. Our research aim is a sound automated theorem prover, built in Isabelle/UTP, for diagrammatic descriptions of reactive, timed, probabilistic controllers for robotics and autonomous systems. We note that it is not straightforward to take the informal proof outlines in [24] and use them directly in a mechanical theorem prover. In this paper, our objective is to explicate the weakest completion semantic technique and in doing so, to explore how to mechanise it.

4   Weakest Preconditions and Prespecifications

In this section, we review Dijkstra's weakest precondition predicate transformer [8] and its generalisation, the weakest prespecification [30].

A typical stage in program development is to prove that a program meets its specification. Schematically, this is a problem in three variables: the program and its specification, which is a precondition and a postcondition. The weakest precondition calculus fixes two of these variables, the program and the postcondition, and calculates the third, the precondition, as an extreme value.

$$
\begin{aligned}
& [\, P \Rightarrow s \Rightarrow q' \,] \\
=\ & [\, s \Rightarrow P \Rightarrow q' \,] \\
=\ & [\, s \Rightarrow \forall\, v' \bullet P \Rightarrow q' \,] \\
=\ & [\, s \Rightarrow \neg\ \exists\, v' \bullet P \wedge \neg\, q' \,] \\
=\ & [\, s \Rightarrow \neg\ \exists\, v_0 \bullet P[v_0/v'] \wedge \neg\, q_0 \,] \\
=\ & [\, s \Rightarrow \neg\, (P \ ;\ \neg\, q) \,]
\end{aligned}
$$

(This derivation is a small variation on that in [31, Chap. 2].) UTP's relational calculus is alphabetised: names are an important part of the meaning. Where we think that it might help, we have emphasised which names occur in each predicate by using parameters. This also streamlines substitution.

Formally, given program P and postcondition q, the problem is to find the weakest precondition s (in terms of P and q) such that P refines $(s \Rightarrow q')$. P refines S, written $P \sqsupseteq S$, just in case $\forall\, v, v' \bullet P \Rightarrow S$, where v and v' denote the before and after states. In UTP, universal closure over an alphabet is abbreviated by brackets, so refinement is defined as $[\, P \Rightarrow S \,]$.

So the predicate $\neg\, (P \ ;\ \neg\, q)$ is the weakest precondition for execution of P to guarantee postcondition q (written as P **wp** q). Here, $P \ ;\ Q$ is the relational composition of P and Q, [31, Chap. 2] defined by $P(s, t') \ ;\ Q(t, u') = \exists\, t_0 \bullet P(s, t_0) \wedge Q(t_0, u')$. Our minor generalisation accounts for its use with non-homogeneous relations later in the paper. Note that there is a modality here, between necessity and possibility. Compare the definition of weakest precondition with its dual, the conjugate weakest precondition [62]: $P\ \overline{\textbf{wp}}\ q = \neg\, (P\ \textbf{wp}\ \neg\, q)$.

$$
\begin{aligned}
& P\ \overline{\textbf{wp}}\ q \\
=\ & \neg\, (P\ \textbf{wp}\ \neg\, q) \\
=\ & \neg\,\neg\, (P \ ;\ \neg\,\neg\, q)
\end{aligned}
$$

$$= \quad P \; ; \; q$$
$$= \quad \exists v' \bullet P \wedge q'$$

During the derivation of weakest precondition, we see that $(P \text{ wp } q) = \forall v' \bullet P \Rightarrow q'$. This has universal force: every final state v' of the program P must satisfy q. Its conjugate has existential force: some execution of P satisfies q.

Now we move on to a generalisation of weakest precondition: the weakest prespecification. First, we define relational converse $P^{\smile}(s, t') = P(s', t)$. For example, the converse of an assignment is calculated as follows:

$$(x := x + 1)^{\smile}$$
$$= \quad (x' = x + 1)^{\smile}$$
$$= \quad (x = x' + 1)$$
$$= \quad (x' = x - 1)$$
$$= \quad x := x - 1$$

Weakest prespecifications generalise weakest preconditions from conditions to relations: given specifications Y and K, find the weakest specification X (in terms of Y and K), such that Y is refined by $X \; ; \; K$. We proceed in a similar way to our previous calculation for the weakest precondition: first, isolate X on the stronger side of the refinement relation, so that we can conclude we have a weakest solution; then rewrite the other side of the relation so that we can use the definition of sequential composition. Our derivation is (as far as we know) novel in the literature. There is a strong analogy between the weakest precondition and weakest prespecification predicate transformers; see Appendix A for further motivation.

$$X \; ; \; K \sqsupseteq Y$$
$$= \quad \{ \text{law of refinement: } (P \; ; \; Q \sqsupseteq R) = (P[x_0/x'] \wedge Q[x_0/x] \sqsupseteq R) \}$$
$$X[s_0/s'] \wedge K[s_0/s] \sqsupseteq Y$$
$$= \quad \{ \text{law of refinement: } (P \wedge Q \sqsupseteq R) = (P \sqsupseteq Q \Rightarrow R) \}$$
$$X[s_0/s'] \sqsupseteq K[s_0/s] \Rightarrow Y$$
$$= \quad \{ \text{change of variables: } s_0, s' \mapsto s', s_0 \}$$
$$X \sqsupseteq K[s', s_0/s, s'] \Rightarrow Y[s_0/s']$$
$$= \quad \{ \text{propositional calculus: contraposition} \}$$
$$X \sqsupseteq \neg Y[s_0/s'] \Rightarrow \neg K[s', s_0/s, s']$$
$$= \quad \{ \text{definition of converse} \}$$
$$X \sqsupseteq \neg Y[s_0/s'] \Rightarrow \neg K^{\smile}[s_0/s]$$
$$= \quad \{ \text{predicate calculus: narrow scope of } s_0 \}$$
$$X \sqsupseteq \forall s_0 \bullet \neg Y[s_0/s'] \Rightarrow \neg K^{\smile}[s_0/s]$$
$$= \quad \{ \text{predicate calculus: De Morgan} \}$$
$$X \sqsupseteq \neg \exists s_0 \bullet \neg Y[s_0/s'] \wedge K^{\smile}[s_0/s]$$

$=$ { definition sequential composition }

$$X \sqsupseteq \neg (\neg Y \; ; \; K^{\smile})$$

So, X must be at least as strong as $\neg (\neg Y \; ; \; K^{\smile})$. We read this as "The weakest prespecification of K through Y", and denote it by Y/K (the weak inverse of the function $\lambda X \bullet X \; ; \; K$). The weakest prespecification forms one adjoint of a Galois connection, with sequential composition as the other adjoint; that is: $(X \; ; \; K \sqsupseteq Y) = (X \sqsupseteq Y/K)$. We give an example of calculating a leading assignment: we want to implement the assignment $x := 2$ as the sequential composition $(X; \; x := x + 1)$. That is, X is the weakest prespecification of $x := x + 1$ through $x := 2$.

$$
\begin{aligned}
& x := 2/x := x + 1 \\
=\; & \neg (\neg x := 2 \; ; \; (x := x + 1)^{\smile}) \\
=\; & \neg (x' \neq 2 \; ; \; (x' = x + 1)^{\smile}) \\
=\; & \neg (x' \neq 2 \; ; \; x = x' + 1) \\
=\; & \neg (\exists x_0 \bullet x_0 \neq 2 \wedge x_0 = x' + 1) \\
=\; & \neg (x' + 1 \neq 2) \\
=\; & x' = 1 \\
=\; & x := 1
\end{aligned}
$$

5   Weakest Completion Semantics

We now turn to weakest completion semantics [24], where we lift standard designs to probabilistic designs. Our objective is to give semantics to a nondeterministic probabilistic programming language that is consistent with a similar standard programming language: the only difference being the presence or absence of a probabilistic choice operator. Consistency is important to make sure that programming intuitions, development techniques, and proof methods can be carried over, as far as possible, from the standard language to the probabilistic one.

One way to achieve consistency is to extend the standard semantics to the probabilistic one in a controlled way. He et al.'s work [24] develops a semantic method to extend theories of programming automatically, as far as possible. Their method is to make only a few explicit assumptions and then generate a semantics by following a set of principles. They have applied their technique to two semantics for the nondeterministic programming language: a relational semantics and a predicate-transformer one.

He et al. propose the following procedure:

1. Start from the semantics for the nondeterministic programming language.
2. Propose a probabilistic semantic domain.

3. Propose a mapping from the probabilistic semantics to the standard semantics to relate computations of probabilistic programs to computations of standard programs.
4. Use this mapping to induce automatically an embedding of programs over the standard semantics: the technique is to consider the weakest completion of a sub-commuting diagram expressing refinement, rather than equality.
5. Determine its defining algebraic characteristics of the new language.

6   Probabilistic Programs

Our standard and our probabilistic programming languages have identical syntax, except that the latter has the addition of a probabilistic choice operator: $P \oplus_r Q$. This is a choice between P with probability r, and Q with probability $1 - r$. The syntax of this language is given in

$$
\begin{array}{llr}
P & ::= & \bot & \text{abort} \\
& | & \mathbb{I} & \text{skip} \\
& | & x := e & \text{assignment} \\
& | & P \lhd b \rhd Q & \text{conditional} \\
& | & P \sqcap Q & \text{nondeterminism} \\
& | & P \oplus_r Q & \text{probabilism} \\
& | & P \; ; \; Q & \text{sequence} \\
& | & \mu X \bullet P(X) & \text{recursion}
\end{array}
$$

This nondeterministic probabilistic language is a suitable target for probabilistic RoboChart [10]. The semantic domain for the language without probabilistic choice is the UTP theory of designs. This theory allows the boolean observation of a program starting (ok) and of it terminating (ok'). A design with precondition $p(s)$ and postcondition $R(s, s')$ is a pair of predicates $(p(s) \vdash R(s, s'))$, which is defined as the single relation $ok \wedge p(s) \Rightarrow ok' \wedge R(s, s')$. This is a statement of total correctness: if the program is started in a state satisfying its precondition, then it will terminate and when it does, its postcondition will be satisfied. The vectors of variables $s, s' : S$ represent the initial and final states of ordinary program variables, which are modelled as mappings from the names of program variables to their values. The UTP semantics for this nondeterministic programming language is well known [31, Chap. 3].

$$
\begin{aligned}
\bot &= (\textbf{false} \vdash \textbf{true}) \\
\mathbb{I} &= (\textbf{true} \vdash s' = s) \\
x := e &= (\textbf{true} \vdash s' = s[e/x]) \\
P \sqcap Q &= P \vee Q \\
P \lhd b \rhd Q &= (b \wedge P) \vee (\neg b \wedge Q) \\
P \; ; \; Q &= \exists \, ok_0, s_0 \bullet P[ok_0, s_0/ok', s'] \wedge Q[ok_0, s_0/ok, s] \\
\mu X \bullet P(X) &= \bigsqcap \{ X \mid X \sqsupseteq P(X) \}
\end{aligned}
$$

Next, we consider the probabilistic semantic domain. Let the state space be S. Let the set of probabilistic distributions over S be the set of total functions to probabilities: $PROB = S \rightarrow [0,1]$. The probabilities in a discrete distribution f sum to 1: $(\sum s : S \bullet f(s)) = 1$, for $f \in PROB$.

A probabilistic design is defined as $p \vdash Q$, where the alphabet of p is $\{s\}$ and the alphabet of Q is $\{s, prob'\}$, for $s \in S$ and $prob' \in PROB$.

The relationship between standard and probabilistic programs is most easily understood as an abstraction from the probabilistic semantic domain: a mapping ρ that forgets the probabilities and replaces them by possibilities. We define ρ as a design with a non-homogeneous alphabet: $\{ok, prob, ok', s'\}$, where ok and ok' design observations about initiation and termination, $prob : PROB$ is a discrete probability distribution, and $s' : S$.

$$\rho \,\,\widehat{=}\,\, (\textbf{true} \vdash prob(s') > 0)$$

This non-homogeneous design is a forgetful function: the *probability* of arriving in state s' is $prob(s')$; this is replaced by the *possibility* of arriving in that state: $prob(s') > 0$.

Note now that $P \,;\, \rho$ is a standard design if P is a probabilistic design.

Using this idea, for probabilistic design P and standard design D, we construct the following sub-commuting diagram

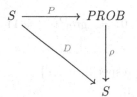

where $P \,;\, \rho \sqsupseteq D$. This is an inequality in three variables, two of which we already know: D and ρ. So, we calculate P using the weakest prespecification of D wrt ρ. The result is the weakest probabilistic design related to the standard design D. We introduce the following definition: for any standard design D, define $\mathcal{K}(D) \,\widehat{=}\, D/\rho$ as its embedding in the probabilistic world.

We need to prove that this embedding really does produce probabilistic designs, which we do in the following theorem. For any subset X of S, define $f(X) = \sum s : X \bullet f(s)$, for any probability distribution function f. Furthermore, for any relation R with alphabet $\{s, s'\}$ (both in S), define $f(R) = f(\{\, s' \mid R \,\})$.[5] If X and Y are disjoint sets then

$$(f(X \cup Y) = 1) = (f(X) = 1 - f(Y))$$

---

[5] Note that if f is a probability distribution function, then lifting f from states to a relation on states results in an alphabetised definition: $f(R)$ has s as a free variable (s' is bound by the set comprehension). If we now fix s, then we get the probability sum for the image of s through R. Note that $prob'(R)$ is also an alphabetised expression, this time with alphabet $\{s, prob'\}$. Thus $prob'(R) = 1$, which we encounter next, is a suitable candidate for the postcondition of a probabilistic design.

A corollary is that

$$(f(R) = 1) = (f(\neg R) = 0)$$

Theorem 1 (Embedded standard designs are probabilistic designs).

$$\mathcal{K}(p(s) \vdash R(s, s')) = (p(s) \vdash prob'(R) = 1)$$

Proof. Start by simplifying the definition of $\mathcal{K}$ by pushing the weakest prespecification operator into the postcondition. Note that the law we use requires that the design is **H3** healthy [31, Chap. 3]: its precondition must not mention any variables from the after-state.[6] This assumption is discharged here. Our account of this law is novel, but we do not present it in this paper.

$$\mathcal{K}(p \vdash R)$$
$$= \quad \{\, \text{definition of } \mathcal{K} \,\}$$
$$(p \vdash R)/(\textbf{true} \vdash prob(s') > 0)$$
$$= \quad \left\{ \begin{array}{l} \text{weakest design prespecification,} \\ P \vdash R \text{ is } \textbf{H3} \text{ implies } (P \vdash Q)/(\textbf{true} \vdash R) = (P \vdash Q/R) \end{array} \right\}$$
$$p \vdash R/(prob(s') > 0)$$

Now show that $R/(prob(s') > 0) = (prob'(R) = 1)$

$$R/(prob(s') > 0)$$
$$= \quad \{\, \text{definition weakest prespecification} \,\}$$
$$\neg (\neg R \; ; \; (prob(s') > 0)^{\smile})$$
$$= \quad \{\, \text{converse} \,\}$$
$$\neg (\neg R \; ; \; prob'(s) > 0)$$
$$= \quad \{\, \text{definition sequential composition} \,\}$$
$$\neg (\exists s_0 \bullet \neg R[s_0/s'] \wedge prob'(s_0) > 0)$$
$$= \quad \{\, \text{predicate calculus} \,\}$$
$$\forall s' \bullet \neg R \Rightarrow prob'(s') = 0$$
$$= \quad \{\, \text{property of lifted probability distribution function} \,\}$$
$$prob'(\neg R) = 0$$

---

[6] This subclass of specification contracts is sometimes known as "normal" designs [14, 21]. The theory of reactive designs [6], mentioned on page 7, is not an embedding of normal designs, since a reactive design can mention the after-value of the trace variable in its precondition. To see this, consider the precondition in the reactive design for the CSP process $a \to CHAOS$. This process can diverge, but only after an a-event. The process's precondition records the circumstances under which the process will not diverge: $\neg \, tr \,^\frown \langle a \rangle \leq tr'$. In words: "Don't press the a button, or else we crash!".

$=$ { property of lifted probability distribution function }
$$prob'(R) = 1$$

Our next task is to prove that the embedding is a homomorphism on the structure of standard programs. As a result, most of the algebraic laws that hold in the standard semantic framework remain valid in the probabilistic model. We give two example cases in the proof of the homomorphism: the embedding of assignment (here) and nondeterminism (in Sect. 8).

Lemma 1 (Embedded assignment).

$$\mathcal{K}(x := e) = (\textbf{true} \vdash prob'(s[e/x]) = 1)$$

Proof.

$\qquad \mathcal{K}(x := e)$
$=$ { semantics of standard assignment }
$\qquad \mathcal{K}(\textbf{true} \vdash s' = s[e/x])$
$=$ { theorem 1 }
$\qquad \textbf{true} \vdash prob'(s' = s[e/x]) = 1$
$=$ { function lifted to relation: $prob(R(s, s')) = prob(\{ s' \mid R(s, s') \})$ }
$\qquad \textbf{true} \vdash prob'(\{ s' \mid s' = s[e/x] \}) = 1$
$=$ { function lifted to set: $prob(X) = \sum s : X \bullet prob(s)$ }
$\qquad \textbf{true} \vdash (\sum s : \{ s' \mid s' = s[e/x] \} \bullet prob'(s)) = 1$
$=$ { set one-point rule: $\{ x \mid x = e \} = \{ e \}$ }
$\qquad \textbf{true} \vdash (\sum s : \{ s[e/x] \} \bullet prob'(s)) = 1$
$=$ { singleton sum: $(\sum x : \{ e \}) = e$ }
$\qquad \textbf{true} \vdash prob'(s[e/x]) = 1$

In the next section, we consider how to combine probability distributions in order to support probabilistic and nondeterministic choice operators.

7   Probabilistic Choice and Combining Distributions

We start with a motivating example of combining probability distributions: expressing multiway probabilistic choice as a combination of binary probabilistic choices. This leads us to propose a semantics for probabilistic choice in the spirit of UTP's parallel-by-merge operator. We consider how to decompose a probability distribution into two distributions combined by probabilistic choice. This leads to two projection functions, one for each operand. We conclude the section with three lemmas that will be used in the case for nondeterministic choice in the proof of $\mathcal{K}$ being a homomorphism. These lemmas provide witnesses for the decomposition required.

Consider a multiway probabilistic choice, as found in the Reactive Modules formalism, [1] used by the probabilistic model checker PRISM [35]:

$$\alpha : (s := 0) \; + \; (1 - (\alpha + \beta)) : (s := 1) \; + \; \beta : (s := 2)$$

Here, each assignment is labelled by a probability and these probabilities sum to 1. How can we express this using binary probabilistic choice? One simple solution uses two operators:

$$(s := 0 \oplus_{\alpha/(1-\beta)} s := 1) \oplus_{1-\beta} s := 2$$

To show that this is a solution, note that the assignment $s := 0$ is chosen with probability $(1 - \beta) \times (\alpha/(1 - \beta)) = \alpha$; $s := 2$ is chosen with probability β; and $s := 1$ must be chosen with the remaining probability, which is $1 - (\alpha + \beta)$. A slightly more complicated solution uses three operators:

$$(s := 0 \oplus_{\alpha+\beta} s := 1) \oplus_{\alpha/(\alpha+\beta)} (s := 1 \oplus_{1-(\alpha+\beta)} s := 2)$$

Analysing probabilities once more gives us $(\alpha/(\alpha + \beta)) \times (\alpha + \beta)$ for $s := 0$; $(1 - (\alpha/(\alpha + \beta))) \times (\alpha + \beta)$ for $s := 2$; and $1 - (\alpha + \beta)$ for $s := 1$. Simple arithmetic proves that we got this right.

These examples show how distributions are combined as we move the binary operator to its multi-way cousin. In the first example, we are combining the following two distributions:[7]

$$0.prob = \{(s = 0) \mapsto \alpha/(1 - \beta), (s = 1) \mapsto 1 - (\alpha/(1 - \beta))\}$$
$$1.prob = \{(s = 2) \mapsto 1\}$$

and we are combining them in the ratio given by the outermost choice operator: $1 - \beta$:

$$
\begin{aligned}
& prob' \\
= \;\; & (1 - \beta) \times 0.prob + (1 - (1 - \beta)) \times 1.prob \\
= \;\; & (1 - \beta) \times 0.prob + \beta \times 1.prob \\
= \;\; & (1 - \beta) \times \{(s = 0) \mapsto \alpha/(1 - \beta), (s = 1) \mapsto 1 - (\alpha/(1 - \beta))\} \\
& + \beta \times \{(s = 2) \mapsto 1\} \\
= \;\; & \{(s = 0) \mapsto (1 - \beta) \times (\alpha/(1 - \beta)), (s = 1) \mapsto (1 - \beta) \times (1 - (\alpha/(1 - \beta)))\} \\
& + \{(s = 2) \mapsto \beta \times 1\} \\
= \;\; & \{(s = 0) \mapsto \alpha, (s = 1) \mapsto 1 - (\alpha + \beta), (s = 2) \mapsto \beta\}
\end{aligned}
$$

To formalise this, define the merge of two distributions, $0.prob$ and $1.prob$, to form distribution $prob'$ as: $M_r = (prob' = r \times 0.prob + (1 - r) \times 1.prob)$, for

---

[7] The notation $0.prob$ and $1.prob$ come from the separating simulation operator in UTP's parallel-by-merge [31, Sect. 7.2], which is being used here to combine probability distributions.

some probability ratio r. We use this in the definition of an operator inspired by UTP's parallel-by-merge [31, Chap. 7] to combine the probability distributions described by two postconditions:

$$P(prob') \parallel_{M_r} Q(prob') = P(0.prob') \wedge Q(1.prob') \; ; \; M_r$$

This operator may be applied equally well to a design, rather than an individual postcondition, without any confusion.

With this operator, we now have a semantics for probabilistic choice:

$$P \oplus_r Q = P \parallel_{M_r} Q$$

The meaning of probabilistic choice is clearly compositional: if we have the meaning of P and Q, then we can find the meaning of $P \oplus_r Q$. But we can also think about the decomposition of a probabilistic program into the probabilistic choice between two subprograms. Suppose that we have two sets of states A and B, such that $A \cup B = S$ and a probabilistic ratio $0 < r < 1$ (to ensure $1/r$ and $1/(1-r)$ are well defined).[8] In this case we can unravel the merge of two distributions if $0.prob(A) = 1$ and $1.prob(B) = 1$. To do this, we define the projections.[9]

$$\mathcal{F}(prob', A, B, r) = (1/r) \times ((A \setminus B) \lhd prob') + ((A \cap B) \lhd prob')$$
$$\mathcal{G}(prob', A, B, r) = (1/(1-r)) \times ((B \setminus A) \lhd prob') + ((A \cap B) \lhd prob')$$

For $\mathcal{F}(prob', A, B, r)$ to be a distribution, we need its domain to sum to unity; that is, $\mathcal{F}(prob', A, B, r)(A) = 1$. These projections satisfy our merge predicate, and in that sense provide a joint witness.

Lemma 2 (Merge witnesses). *For* $0 < r < 1$, $\mathcal{F}(prob', A, B, r)(A) = 1$, *and* $\mathcal{G}(prob', A, B, r)(B) = 1$,

$$M_r[\mathcal{F}(prob', A, B, r), \mathcal{G}(prob', A, B, r)/0.prob, 1.prob]$$

Proof.

$$M_r[\mathcal{F}(prob', A, B, r), \mathcal{G}(prob', A, B, r)/0.prob, 1.prob]$$
$=$ { definition of M_r }
$$\left(prob' = r \times 0.prob + (1-r) \times 1.prob \right) \left[\begin{matrix} \mathcal{F}(prob', A, B, r)/0.prob \\ \mathcal{G}(prob', A, B, r)/1.prob \end{matrix} \right]$$
$=$ { substitution }
$$prob' = r \times \mathcal{F}(prob', A, B, r) + (1-r) \times \mathcal{G}(prob', A, B, r)$$
$=$ { definitions of $\mathcal{F}$ and $\mathcal{G}$ }

---

[8] This case analysis is present in [24], although its purpose is not explained there).

[9] The expression $S \lhd R$ is Z's domain restriction operator [53, p. 98]: the domain restriction $S \lhd R$ of a relation R to a set S relates x to y if and only if R relates x to y and x is a member of S.

$$prob' = r \times ((1/r) \times ((A \setminus B) \lhd prob') + ((A \cap B) \lhd prob'))$$
$$+ (1 - r) \times ((1/(1 - r)) \times ((B \setminus A) \lhd prob') + ((A \cap B) \lhd prob'))$$

$=$ { function scaling: $x \times (f + g) = x \times f + x \times g$, arithmetic }

$$prob' = ((A \setminus B) \lhd prob') + r \times ((A \cap B) \lhd prob')$$
$$+ ((B \setminus A) \lhd prob') + (1 - r) \times ((A \cap B) \lhd prob')$$

$=$ { function scaling: $(x + y) \times f = x \times f + y \times f$, arithmetic }

$$prob' = ((A \setminus B) \lhd prob') + ((A \cap B) \lhd prob') + ((B \setminus A) \lhd prob')$$

$=$ { function addition: $X \cap Y = \varnothing \Rightarrow (X \cup Y) \lhd f = (X \lhd f) + (Y \lhd f)$ }

$$prob' = ((A \setminus B) \cup (A \cap B) \cup (B \setminus A)) \lhd prob'$$

$=$ { assumption: $A \cup B = S$ }

$true$

Now we state two lemmas that ensure that our two projections are probability distributions that sum to unity.

Lemma 3 (Total witness 1). *Let* $p(A \setminus B) = \alpha$ *and* $p(A \cap B) = 1 - (\alpha + \beta)$; *then*

$$\mathcal{F}(p, A, B, \alpha/(\alpha + \beta))(A) = 1$$

Proof.

$\mathcal{F}(p, A, B, \alpha/(\alpha + \beta))(A)$

$=$ { definition $\mathcal{F}$ }

$((1/(\alpha/(\alpha + \beta))) \times ((A \setminus B) \lhd p) + (A \cap B) \lhd p)(A)$

$=$ { arithmetic, function scaling: $(f + g)(X) = f(X) + g(X)$ }

$((\alpha + \beta)/\alpha) \times (A \setminus B) \lhd p(A) + (A \cap B) \lhd p(A)$

$=$ { functions: $X \subseteq Y \Rightarrow X \lhd f(Y) = f(X)$ }

$((\alpha + \beta)/\alpha) \times p(A \setminus B) + p(A \cap B)$

$=$ { assumptions: $p(A \setminus B) = \alpha$ and $p(A \cap B) = 1 - (\alpha + \beta)$ }

$((\alpha + \beta)/\alpha) \times \alpha + 1 - (\alpha + \beta)$

$=$ { arithmetic }

$\alpha + \beta + 1 - \alpha - \beta$

$=$ { arithmetic }

1

Lemma 4 (Total witness 2). *Let* $p(B \setminus A) = \beta$ *and* $p(A \cap B) = 1 - (\alpha + \beta)$; *then*

$$\mathcal{G}(p, A, B, \alpha/(\alpha + \beta))(B) = 1$$

Proof. Similar to Lemma 3.

The main result that we want to present in this paper is stated and proved in the next section.

8   Nondeterministic Choice

In this section, we prove the case for nondeterministic choice in the homomorphism theorem. Nondeterministic choice can be used in the top-down development of a program to abstract from detail, including specific details of a probabilistic choice. So $\mathcal{K}$ should distribute through nondeterministic choice in the following way:

$$\mathcal{K}(D_0 \sqcap D_1) \ = \ \exists\, r : [0,1] \bullet (\mathcal{K}(D_0) \parallel_{M_r} \mathcal{K}(D_1))$$

Refinement to a particular probabilistic choice $\oplus_\alpha$ would then follow by strengthening the result, choosing α for r. In this section, we prove one half of this result, omitting the other half only because we lack space.

The next lemma simplifies the embedding of nondeterministic choice.

Lemma 5 (Embedded nondeterministic choice).

$$\mathcal{K}((p_0 \vdash Q_0) \sqcap (p_1 \vdash Q_1)) \ = \ (p_0 \wedge p_1 \vdash prob'(Q_0 \vee Q_1) = 1)$$

Proof.

$$\mathcal{K}((p_0 \vdash Q_0) \sqcap (p_1 \vdash Q_1))$$

$$= \left\{ \begin{array}{l} \text{designs closed under nondeterministic choice:} \\ ((p_0 \vdash Q_0) \sqcap (p_1 \vdash Q_1)) = (p_0 \wedge p_1 \vdash Q_0 \vee Q_1) \end{array} \right\}$$

$$\mathcal{K}(p_0 \wedge p_1 \vdash Q_0 \vee Q_1)$$

$$= \{ \text{definition of } \mathcal{K}:\ \mathcal{K}(p \vdash Q) = (p \vdash prob'(Q) = 1) \}$$

$$p_0 \wedge p_1 \vdash prob'(Q_0 \vee Q_1) = 1$$

Now we show half of our result: that the embedding is a weakening homomorphism for nondeterministic choice. This means that as $\mathcal{K}$ distributes through nondeterminism, it produces a weaker predicate.

Theorem 2 (Nondeterminism embedding weakening).

$$\mathcal{K}((p_0 \vdash Q_0) \sqcap (p_1 \vdash Q_1)) \sqsupseteq \exists\, r : [0,1] \bullet (\mathcal{K}(p_0 \vdash Q_0) \parallel_{M_r} \mathcal{K}(p_1 \vdash Q_1))$$

Proof.

$$\mathcal{K}((p_0 \vdash Q_0) \sqcap (p_1 \vdash Q_1))$$

$$= \left\{ \begin{array}{l} \text{lemma 5: embedded nondeterministic choice:} \\ \mathcal{K}((p_0 \vdash Q_0) \sqcap (p_1 \vdash Q_1)) \ = \ p_0 \wedge p_1 \vdash prob'(Q_0 \vee Q_1) = 1 \end{array} \right\}$$

$$p_0 \wedge p_1 \vdash prob'(Q_0 \vee Q_1) = 1$$

$$\Rightarrow \left\{ \begin{array}{l} \alpha := p(Q_0 \setminus Q_1) \wedge \beta := p(Q_1 \setminus Q_0) \Rightarrow p(Q_0 \cap Q_1) = 1 - (\alpha + \beta) \\ \text{lemma 3: total witness 1, lemma 4: total witness 2} \end{array} \right\}$$

$$p_0 \wedge p_1 \vdash \mathcal{F}(prob', Q_0, Q_1, \alpha/(\alpha + \beta))(Q_0) = 1$$
$$\wedge \, \mathcal{G}(prob', Q_0, Q_1, \alpha/(\alpha + \beta))(Q_1) = 1$$

$= \{ \text{ lemma 2: merge witnesses } \}$

$$p_0 \wedge p_1 \vdash \mathcal{F}(prob', Q_0, Q_1, \alpha/(\alpha + \beta))(Q_0) = 1$$
$$\wedge \, \mathcal{G}(prob', Q_0, Q_1, \alpha/(\alpha + \beta))(Q_1) = 1$$
$$\wedge \, M_{\alpha/(\alpha+\beta)} \begin{bmatrix} \mathcal{F}(prob', A, B, \alpha/(\alpha + \beta))/0.prob \\ \mathcal{G}(prob', A, B, \alpha/(\alpha + \beta))/1.prob \end{bmatrix}$$

$= \left\{ \begin{array}{l} \text{existential introduction: } r := \alpha/(\alpha + \beta), \\ 0.prob_0 := \mathcal{F}(prob', A, B, r), \, 1.prob_0 := \mathcal{G}(prob', A, B, r) \end{array} \right\}$

$$p_0 \wedge p_1 \vdash \exists\, r : [0, 1]; \; 0.prob_0, 1.prob_0 : PROB \bullet$$
$$0.prob_0(Q_0) = 1 \wedge 1.prob_0(Q_1) = 1$$
$$\wedge \, M_r[0.prob_0, 1.prob_0/0.prob, 1.prob]$$

$= \{ \text{ sequential composition } \}$

$$\exists\, r : [0, 1] \bullet (p_0 \wedge p_1 \vdash 0.prob'(Q_0) = 1 \wedge 1.prob(Q_1) = 1 \; ; \; M_r)$$

$= \{ \text{ definition merge operator } \}$

$$\exists\, r : [0, 1] \bullet (p_0 \vdash prob'(Q_0) = 1) \,\|_{M_r} (p_1 \vdash prob'(Q_1) = 1)$$

$= \{ \text{ definition } \mathcal{K} \}$

$$\exists\, r : [0, 1] \bullet (\mathcal{K}(p_0 \vdash Q_0) \,\|_{M_r} \mathcal{K}(p_1 \vdash Q_1))$$

We omit the (easier) proof that the embedding is a strengthening homomorphism for nondeterministic choice: as $\mathcal{K}$ distributes through nondeterminism we obtain a stronger predicate.

This concludes our presentation of the semantics for the nondeterministic probabilistic programming language that serves as the textual version of RoboChart diagrams with discrete probabilistic behaviour. We have described the semantic domain and an embedding function from standard programs to probabilistic ones. We have shown just two cases for the proof that the embedding is a homomorphism. This has guided the definition of individual program operators. For example, we have

$$\mathcal{K}(D_0 \sqcap D_1) = \mathcal{K}(D_0) \vee \mathcal{K}(D_1) \vee \bigvee_{0 < r < 1}(\mathcal{K}(D_0) \,\|_{M_r} \mathcal{K}(D_1))$$

This definition is supported by Theorem 2 and a matching proof for the strengthening homomorphism (omitted in this paper). The proof identified the need for the two special cases in the semantics of nondeterminism: $r = 0$ and $r = 1$.

9   Related Work

Jansen et al. propose a probabilistic extension to UML [33,34]. They add to UML's basic Statecharts a probabilistic choice node whose out-edges are annotated with probabilities. They identify interferences between Statechart transition priorities and the order of resolving nondeterministic and probabilistic choice. Verification is performed using the PRISM probabilistic model checker, with the probabilistic logic PCTL specifying properties over Statecharts. They describe the operational semantics of step execution. This is then embedded in a finite Markov Decision Process specified as a probabilistic Kripke system.

Nokovic and Sekerinski [43] propose pCharts, another variation on Statecharts, but extended with timed transitions, probabilistic transitions, costs and rewards, and state invariants. They present a translation scheme from untimed pCharts to Markov Decision Processes (MDPs), from timed pCharts to probabilistic timed automata (PTA), and from pCharts to executable C code. Everything is implemented in the pState tool. MDPs are used to verify probabilistic and nondeterministic behaviour. PTAs are used to verify additional real-time constraints, such as the maximum or minimum probability of reaching a state within a given time and the maximum expected time to reach that state (its deadline). pCharts can be augmented with quantitative information for costs and rewards for both transitions and states: priced PTAs. This permits analysis of the maximum or minimum expected time before a transition takes place, or the number of expected steps to reach a particular state. Translation rules deal with hierarchy and orthogonality.

Both Jansen's and Nokovic's work is similar to He et al.'s [24], and therefore ours, in constructing a conservative extension of standard Statecharts. Both of them go further in dealing with hierarchy and orthogonality. This differs from our work in several ways. We focus on producing a semantics that can be combined with other UTP theories. Both Jansen and Nokovic focus on model checking, and therefore have a closed-world assumption and restrict variables to bound integers. We are interested in both model checking and theorem proving.

In 2004, Goldsmith reported an experiment [20] to extend the input language for FDR2 to accept a probabilistic choice construct with added functionality, to produce models suitable for analysis by PRISM [35]. Goldsmith describes some encouraging results, but also warns about various drawbacks in the work: the loss of regularity in code emitted from FDR2 that would lead to PRISM exploiting symmetries in its model checking; and that the transformation scheme does not support CSP's full failures-divergences model. The probabilistic functionality in FDR2 was lost when development moved to FDR3 in 2012 and remained lost with the move to FDR4 in 2017.

Mota et al. [10] rediscovered the functionality in FDR2 (as well as legacy copies of the tool) in their work on analysing probability in RoboChart. They define the semantics of the RoboChart probabilistic choice operator in terms of CSP's probabilistic operator. They show how this augmented CSP semantics for RoboChart can be translated into the PRISM's Reactive Modules input language to check stochastic properties of RoboChart.

Zhao et al. [65] describe mapping rules between UML state diagrams and probabilistic Kripke structure semantics. They present an asynchronous parallel language based on discrete time Markov chains. Non-functional properties of systems specified using PCTL, with verification provided by the PRISM model checker. Interactive theorem proving is also supported and linked to experimental results. Interestingly, the mapping rules are provided as a bidirectional transformation.

Zhang et al. [64] address the formal verification of dynamic behaviour of UML diagrams. They automatically verify UML state machine models by translating UML models to the input language of the PAT model checker in such a way as to be transparent for users. They can check safety and liveness properties with fairness assumptions using the PAT model checker [38].

10   Conclusions and Future Work

We have presented an overview of our ongoing work in giving a probabilistic semantics to RoboChart. We have concentrated on the imperative, sequential action language for RoboChart, using the weakest completion semantics approach. The result is a programming theory that can now be combined with other programming paradigms, using UTP's unification techniques explained in Sect. 3. The next step for us is to lift the current semantics into UTP's reactive theory to produce a theory of reactive probabilistic designs.

We have explicated the weakest completion approach, showing how proof outlines in [24] can be turned into near formal proofs suitable for implementation in a theorem prover. In doing this, we spent a surprising amount of time understanding the structure of He et al.'s proof, especially the nondeterminism case for the proof that the embedding function $\mathcal{K}$ is a homomorphism. This led us to investigate the weakest prespecification operator for the design theory in some detail, coming up with what we believe to be a novel derivation of the operator that echoes Hoare and He's derivation of the weakest precondition operator [31]. We observe that a law quoted in the proof of this case in [24] requires a side condition that the design to which it is applied satisfies the **H3** healthiness condition [31, Chap. 3]. This is the case in the proof where it is used, but it does raise an interesting question for our lifting the current semantics to the reactive world. We found a small number of inconsistencies in the proof outlines, but these have not affected the validity of the lemmas and theorems in [24].

Our future work consists of the following:

1. Complete the rest of the proof that $\mathcal{K}$ is a homomorphism (essentially, the Kleisli lifting needed for sequential composition).
2. Implement our proofs in the Isabelle/UTP theorem prover [19].[10]
3. Lift the semantics to the reactive theory.
4. Use our semantics to verify the soundness of a translation from RoboChart to Reactive Modules, so that PRISM can be used to analyse probabilistic RoboCharts.
5. Tackle a range of different examples using both model checking and theorem proving to challenge our work. We have in our sights various probabilistic robotic control algorithms.

---

[10] We have already begun work on the mechanisation of the proofs in Isabelle/UTP. Early indications show that the meticulous detail in the hand-written proofs is very helpful in the mechanisation.

Examples include verifying robot localisation algorithms, such as the Random Sample Consensus algorithm RANSAC that is frequently used in robotic control [11]; providing bounds for the battery life required for coverage using random walks and arena-mapping techniques by autonomous robotic cleaners and searchers; and verifying learning algorithms for robots in uncertain environments.

Acknowledgements. This work was funded under EPSRC grant EP/M025756/1 on A Calculus for Software Engineering of Mobile and Autonomous Robots, Royal Society grant Requirements Modelling for Cyber-Physical Systems, and a Royal Academy of Engineering Chair in Emerging Technologies. We are grateful for very helpful feedback from the reviewers that helped us clarify the exposition of our ideas in this paper (including the explanation of the connection between weakest precondition and weakest prespecification in Appendix A). We have benefited from discussions with Riccardo Bresciani, Andrew Butterfield, Ana Cavalcanti, Tony Hoare, Lydia Hughes, Zhiming Liu, Alvaro Miyazawa, and Augusto Sampaio. We are especially grateful to He Jifeng, Annabelle McIver, and Carroll Morgan for their beautiful ideas. The work in this paper was first presented at the IFIP WG 2.3 (Programming Methodology) meeting in York in February 2019 and at a Royal Society/National Natural Science Foundation of China workshop at Southwest University (Chongqing) in May 2019.

A   Connecting Weakest Preconditions and Prespecifications

Weakest preconditions and prespecifications each arise as the weakest solution of an inequality in three variables. Both have a conjunction on the implementation side. The inequality for the weakest precondition in stated as $P \sqsupseteq s \Rightarrow q$, but this is equivalent to $s \wedge P \sqsupseteq q$ (1). The inequality for the weakest prespecification is stated as $X \; ; \; K \sqsupseteq Y$, but this is equivalent to $X[v_0/v'] \wedge K[v_0/v] \sqsupseteq Y$ (2). The two inequalities have the same essential structure. Hoare & He go further and note as a conjecture that the two predicate transformers are almost identical when the first argument mentions only dashed variables: $r'/K = (K \text{ } \textbf{wp} \text{ } r)'$. The conjecture is easily proved.

$$r'/K$$
$= \quad \{\, \text{dashing a condition: } c' = c[v'/v] \,\}$
$$r[v'/v]/K$$
$= \quad \{\, \text{definition of weakest prespecification} \,\}$
$$\neg\,(\neg\,r[v'/v] \; ; \; K^{\smile})$$
$= \quad \{\, \text{definition of relational converse} \,\}$
$$\neg\,(\neg\,r[v'/v] \; ; \; K[v', v/v, v']))$$
$= \quad \{\, \text{definition of sequential composition} \,\}$
$$\neg\,\exists\,v_0 \bullet \neg\,r[v_0/v] \wedge K[v', v_0/v, v'])$$
$= \quad \{\, \text{propositional calculus} \,\}$

$$\neg \ \exists \, v_0 \bullet K\,[v', v_0/v, v'] \wedge \neg \ r[v_0/v])$$
$$= \ \{\text{dashing a relation: } R' = R[v'/v]\,\}$$
$$\neg \ (\exists \, v_0 \bullet K\,[v, v_0/v, v'] \wedge \neg \ r[v_0/v])'$$
$$= \ \{\text{definition of relational converse}\,\}$$
$$\neg \ (K \ ; \ \neg \ r)'$$
$$= \ \{\text{definition weakest precondition}\,\}$$
$$(K \ \mathbf{wp} \ r)'$$

This result means that the weakest prespecification subsumes the weakest precondition and so could be used to give its definition: $K \ \mathbf{wp} \ r \ \widehat{=} \ (r'/K)[v/v']$.

References

1. Alur, R., Henzinger, T.A.: Reactive modules. Formal Methods Syst. Des. **15**(1), 7–48 (1999)
2. Bousmalis, K.: Closing the simulation-to-reality gap for deep robotic learning (2019). Google AI Blog http://ai.googleblog.com/2017/10/closing-simulation-to-reality-gap-for.html
3. Brunner, S.G., Steinmetz, F., Belder, R., Dömel, A.: RAFCON: a graphical tool for engineering complex, robotic tasks. In: 2016 IEEE/RSJ International Conference on Intelligent Robots and Systems, IROS 2016, Daejeon, South Korea, 9–14 October 2016, pp. 3283–3290 (2016)
4. Cavalcanti, A., Ribeiro, P., Miyazawa, A., Sampaio, A., Filho, M.C., Didier, A.: RoboSim: Reference Manual (2019). www.cs.york.ac.uk/robostar/robosim/robosim-reference.pdf
5. Cavalcanti, A., Sampaio, A., Woodcock, J.: Refinement of actions in Circus. Electr. Notes Theor. Comput. Sci. **70**(3), 132–162 (2002)
6. Cavalcanti, A., Woodcock, J.: A tutorial introduction to CSP in *Unifying Theories of Programming*. In: Cavalcanti, A., Sampaio, A., Woodcock, J. (eds.) PSSE 2004. LNCS, vol. 3167, pp. 220–268. Springer, Heidelberg (2006). https://doi.org/10.1007/11889229_6
7. Dhouib, S., Kchir, S., Stinckwich, S., Ziadi, T., Ziane, M.: RobotML, a domain-specific language to design, simulate and deploy robotic applications. In: Noda, I., Ando, N., Brugali, D., Kuffner, J.J. (eds.) SIMPAR 2012. LNCS (LNAI), vol. 7628, pp. 149–160. Springer, Heidelberg (2012). https://doi.org/10.1007/978-3-642-34327-8_16
8. Dijkstra, E.W.: A Discipline of Programming. Prentice-Hall, Upper Saddle River (1976)
9. FDR: Failures-Divergences Refinement. www.cs.ox.ac.uk/projects/fdr/
10. Conserva Filho, M.S., Marinho, R., Mota, A., Woodcock, J.: Analysing RoboChart with probabilities. In: Massoni, T., Mousavi, M.R. (eds.) SBMF 2018. LNCS, vol. 11254, pp. 198–214. Springer, Cham (2018). https://doi.org/10.1007/978-3-030-03044-5_13
11. Fischler, M.A., Bolles, R.C.: Random sample consensus: a paradigm for model fitting with applications to image analysis and automated cartography. Commun. ACM **24**(6), 381–395 (1981)

12. Fitzgerald, J.S., Gamble, C., Larsen, P.G., Pierce, K., Woodcock, J.: Cyber-physical systems design: Formal foundations, methods and integrated tool chains. In: Gnesi, S., Plat, N. (eds.) 3rd IEEE/ACM FME Workshop on Formal Methods in Software Engineering, FormaliSE 2015, Florence, 18 May 2015, pp. 40–46. IEEE Computer Society (2015)

13. Foster, S., Baxter, J., Cavalcanti, A., Miyazawa, A., Woodcock, J.: Automating verification of state machines with reactive designs and Isabelle/UTP. In: Bae, K., Ölveczky, P.C. (eds.) FACS 2018. LNCS, vol. 11222, pp. 137–155. Springer, Cham (2018). https://doi.org/10.1007/978-3-030-02146-7_7

14. Foster, S., Cavalcanti, A., Canham, S., Woodcock, J., Zeyda, F.: Unifying theories of reactive design contracts. CoRR abs/1712.10233 (2017). arxiv.org/abs/1712.10233

15. Foster, S., Cavalcanti, A., Woodcock, J., Zeyda, F.: Unifying theories of time with generalised reactive processes. Inf. Process. Lett. **135**, 47–52 (2018)

16. Foster, S., Woodcock, J.: Unifying theories of programming in Isabelle. In: Liu, Z., Woodcock, J., Zhu, H. (eds.) Unifying Theories of Programming and Formal Engineering Methods. LNCS, vol. 8050, pp. 109–155. Springer, Heidelberg (2013). https://doi.org/10.1007/978-3-642-39721-9_3

17. Foster, S., Woodcock, J.: Towards verification of cyber-physical systems with UTP and Isabelle/HOL. In: Gibson-Robinson, T., Hopcroft, P., Lazić, R. (eds.) Concurrency, Security, and Puzzles. LNCS, vol. 10160, pp. 39–64. Springer, Cham (2017). https://doi.org/10.1007/978-3-319-51046-0_3

18. Foster, S., Zeyda, F., Nemouchi, Y., Ribeiro, P., Wolff, B.: Isabelle/UTP: mechanised theory engineering for unifying theories of programming. Arch. Formal Proofs (2019)

19. Foster, S., Zeyda, F., Woodcock, J.: Isabelle/UTP: a mechanised theory engineering framework. In: Naumann, D. (ed.) UTP 2014. LNCS, vol. 8963, pp. 21–41. Springer, Cham (2015). https://doi.org/10.1007/978-3-319-14806-9_2

20. Goldsmith, M.: CSP: the best concurrent-system description language in the world–probably! In: Communicating Process Architectures, pp. 227–232 (2004)

21. Guttmann, W., Möller, B.: Normal design algebra. J. Log. Algebr. Program. **79**(2), 144–173 (2010)

22. Harel, D.: Statecharts: a visual formalism for complex systems. Sci. Comput. Program. **8**(3), 231–274 (1987)

23. Harwood, W., Cavalcanti, A., Woodcock, J.: A theory of pointers for the UTP. In: Fitzgerald, J.S., Haxthausen, A.E., Yenigun, H. (eds.) ICTAC 2008. LNCS, vol. 5160, pp. 141–155. Springer, Heidelberg (2008). https://doi.org/10.1007/978-3-540-85762-4_10

24. Jifeng, H., Morgan, C., McIver, A.: Deriving probabilistic semantics via the 'Weakest Completion'. In: Davies, J., Schulte, W., Barnett, M. (eds.) ICFEM 2004. LNCS, vol. 3308, pp. 131–145. Springer, Heidelberg (2004). https://doi.org/10.1007/978-3-540-30482-1_17

25. Hehner, E.C.R.: Predicative programming, part I. Commun. ACM **27**(2), 134–143 (1984)

26. Hehner, E.C.R.: Predicative programming, part II. Commun. ACM **27**(2), 144–151 (1984)

27. Hehner, E.C.R., Gupta, L.E., Malton, A.J.: Predicative methodology. Acta Inf. **23**(5), 487–505 (1986)

28. Hilder, J.A., et al.: Chemical detection using the receptor density algorithm. IEEE Trans. Syst. Man Cybern. Part C **42**(6), 1730–1741 (2012)

29. Hoare, C.A.R.: Programs are predicates. In: FGCS, pp. 211–218 (1992)
30. Hoare, C.A.R., He, J.: The weakest prespecification. Inf. Process. Lett. **24**(2), 127–132 (1987)
31. Hoare, C.A.R., He, J.: Unifying Theories of Programming. Prentice Hall, Upper Saddle River (1998)
32. Jakobi, N., Husbands, P., Harvey, I.: Noise and the reality gap: the use of simulation in evolutionary robotics. In: Morán, F., Moreno, A., Merelo, J.J., Chacón, P. (eds.) ECAL 1995. LNCS, vol. 929, pp. 704–720. Springer, Heidelberg (1995). https://doi.org/10.1007/3-540-59496-5_337
33. Jansen, D.N., Hermanns, H., Katoen, J.-P.: A probabilistic extension of UML statecharts. In: Damm, W., Olderog, E.-R. (eds.) FTRTFT 2002. LNCS, vol. 2469, pp. 355 374. Springer, Heidelberg (2002). https://doi.org/10.1007/3-540-45739-9_21
34. Jansen, D.: Extensions of Statecharts with probability, time, and stochastic timing. Ph.D. thesis, University of Twente (2003)
35. Kwiatkowska, M.Z., Norman, G., Parker, D.: PRISM: probabilistic symbolic model checker. In: Field, T., Harrison, P.G., Bradley, J., Harder, U. (eds.) TOOLS 2002. LNCS, vol. 2324, pp. 200–204. Springer, Heidelberg (2002). https://doi.org/10.1007/3-540-46029-2_13
36. Larsen, P.G., et al.: Integrated tool chain for model-based design of cyber-physical systems: the INTO-CPS project. In: 2016 2nd International Workshop on Modelling, Analysis, and Control of Complex CPS, CPS Data 2016, Vienna, 11 April 2016, pp. 1–6. IEEE Computer Society (2016)
37. Lee, E.A., Seshia, S.A.: Introduction to Embedded Systems: A Cyber-Physical Systems Approach, 2nd edn. The MIT Press, Cambridge (2016)
38. Liu, Y., Sun, J., Dong, J.S.: PAT 3: an extensible architecture for building multi-domain model checkers. In: Dohi, T., Cukic, B. (eds.) IEEE 22nd International Symposium on Software Reliability Engineering, ISSRE 2011, Hiroshima, 29 November–2 December 2011, pp. 190–199. IEEE Computer Society (2011)
39. Miyazawa, A.: RoboTool: RoboChart Tool Manual. University of York (2018). http://tinyurl.com/RoboTool-Manual
40. Miyazawa, A., Ribeiro, P., Li, W., Cavalcanti, A., Timmis, J.: Automatic property checking of robotic applications. In: 2017 IEEE/RSJ International Conference on Intelligent Robots and Systems, IROS 2017, Vancouver, 24–28 September 2017, pp. 3869–3876 (2017)
41. Miyazawa, A., Ribeiro, P., Li, W., Cavalcanti, A., Timmis, J., Woodcock, J.: RoboChart: modelling and verification of the functional behaviour of robotic applications. Softw. Syst. Model. **18**, 3097–3149 (2019)
42. Nipkow, T., Wenzel, M., Paulson, L.C.: Isabelle/HOL—A Proof Assistant for Higher-Order Logic. LNCS, vol. 2283. Springer, Heidelberg (2002). https://doi.org/10.1007/3-540-45949-9
43. Nokovic, B., Sekerinski, E.: Verification and code generation for timed transitions in pCharts. In: Desai, B.C. (ed.) International C* Conference on Computer Science & Software Engineering, C3S2E 2014, Montreal, 3–5 August 2014, pp. 3:1–3:10. ACM (2014)
44. Object Management Group: OMG Unified Modeling Language (OMG UML), superstructure, version 2.4.1
45. Oliveira, M., Cavalcanti, A., Woodcock, J.: A denotational semantics for Circus. Electr. Notes Theor. Comput. Sci. **187**, 107–123 (2007)
46. Oliveira, M., Cavalcanti, A., Woodcock, J.: A UTP semantics for *Circus*. Formal Asp. Comput. **21**(1–2), 3–32 (2009)

47. Pembeci, I., Nilsson, H., Hager, G.D.: Functional reactive robotics: an exercise in principled integration of domain-specific languages. In: Proceedings of the 4th International ACM SIGPLAN Conference on Principles and Practice of Declarative Programming, 6–8 October 2002, Pittsburgh (Affiliated with PLI 2002), pp. 168–179 (2002)

48. Ribeiro, P., Miyazawa, A., Li, W., Cavalcanti, A., Timmis, J.: Modelling and verification of timed robotic controllers. In: Polikarpova, N., Schneider, S. (eds.) IFM 2017. LNCS, vol. 10510, pp. 18–33. Springer, Cham (2017). https://doi.org/10.1007/978-3-319-66845-1_2

49. RoboCalc. www.cs.york.ac.uk/circus/RoboCalc

50. RoboCalc Project: The foraging robot example. University of York (2019). http://tinyurl.com/y4h9aq2l

51. Roscoe, A.W.: On the expressive power of CSP refinement. Formal Asp. Comput. **17**(2), 93–112 (2005)

52. Roscoe, A.W.: Understanding Concurrent Systems. Texts in Computer Science. Springer, Heidelberg (2010). https://doi.org/10.1007/978-1-84882-258-0

53. Spivey, J.: The Z Notation: A Reference Manual, 2nd edn. Prentice-Hall, Upper Saddle River (1989)

54. V-REP: Virtual Robot Experimentation Platform, User Manual, Version 3.6.1. www.coppeliarobotics.com/helpFiles/en/importExport.htm

55. Wächter, M., Ottenhaus, S., Kröhnert, M., Vahrenkamp, N., Asfour, T.: The ArmarX Statechart concept: graphical programming of robot behavior. Front. Robot. AI **3**, 33 (2016)

56. Webots: Reference Manual, Rel. R2019a. www.cyberbotics.com/doc/reference/

57. Winfield, A.F.T.: Foraging robots. In: Meyers, R.A. (ed.) Encyclopedia of Complexity and Systems Science, pp. 3682–3700. Springer, Heidelberg (2009). https://doi.org/10.1007/978-0-387-30440-3_217

58. Woodcock, J.: Engineering UToPiA: formal semantics for CML. In: Jones, C., Pihlajasaari, P., Sun, J. (eds.) FM 2014. LNCS, vol. 8442, pp. 22–41. Springer, Cham (2014). https://doi.org/10.1007/978-3-319-06410-9_3

59. Woodcock, J., Cavalcanti, A.: A tutorial introduction to designs in unifying theories of programming. In: Boiten, E.A., Derrick, J., Smith, G. (eds.) IFM 2004. LNCS, vol. 2999, pp. 40–66. Springer, Heidelberg (2004). https://doi.org/10.1007/978-3-540-24756-2_4

60. Woodcock, J., Foster, S.: UTP by example: designs. In: Bowen, J.P., Liu, Z., Zhang, Z. (eds.) SETSS 2016. LNCS, vol. 10215, pp. 16–50. Springer, Cham (2017). https://doi.org/10.1007/978-3-319-56841-6_2

61. Woodcock, J., Foster, S., Butterfield, A.: Heterogeneous semantics and unifying theories. In: Margaria, T., Steffen, B. (eds.) ISoLA 2016. LNCS, vol. 9952, pp. 374–394. Springer, Cham (2016). https://doi.org/10.1007/978-3-319-47166-2_26

62. Woodcock, J.C.P., Morgan, C.: Refinement of state-based concurrent systems. In: Bjørner, D., Hoare, C.A.R., Langmaack, H. (eds.) VDM 1990. LNCS, vol. 428, pp. 340–351. Springer, Heidelberg (1990). https://doi.org/10.1007/3-540-52513-0_18

63. Zave, P., Jackson, M.: Conjunction as composition. ACM Trans. Softw. Eng. Methodol. **2**(4), 379–411 (1993)

64. Zhang, S.J., Liu, Y.: An automatic approach to model checking UML state machines. In: Fourth International Conference on Secure Software Integration and Reliability Improvement, SSIRI 2010, Singapore, 9–11 June 2010, pp. 1–6. IEEE Computer Society (2010)

65. Zhao, Y., Yang, Z., Xie, J., Liu, Q.: Quantitative analysis of system based on extended UML state diagrams and probabilistic model checking. JSW **5**(7), 793–800 (2010)

Hybrid Models

Unified Graphical Co-modelling
of Cyber-Physical Systems Using AADL
and Simulink/Stateflow

Haolan Zhan[1,2], Qianqian Lin[1,2], Shuling Wang[1], Jean-Pierre Talpin[3], Xiong Xu[1],
and Naijun Zhan[1,2(✉)]

[1] State Key Laboratory of Computer Science, Institute of Software,
CAS, Beijing, China
{zhanhl,linqq,wangsl,xux,znj}@ios.ac.cn
[2] University of Chinese Academy of Sciences, Beijing, China
[3] Institut National de Recherche en Informatique et en Automatique (INRIA), Rennes, France
jean-pierre.talpin@inria.fr

Abstract. The efficient design of safety-critical embedded systems involves, at
least, the three modelling aspects common to all cyber-physical systems (CPSs):
functionalities, *physics* and *architectures*. Existing modelling formalisms cannot
provide strong support to take all of these three dimensions into account uniformly, e.g., AADL is a precise formalism for modelling architecture and prototyping hardware platforms, but it is weak for modelling physical and software
behaviours and their interaction. By contrast, Simulink/Stateflow is strong for
modelling physical and software behaviour and their interaction, but weak for
modelling architecture and hardware platforms. To address this issue, we consider
the combination of AADL and Simulink/Stateflow, two widely used graphical
modelling formalisms for CPS design in industry. This combination provides a
unified graphical co-modelling formalism supporting the design of CPSs from all
three software, hardware and physics perspectives uniformly. This paper focuses
on the required concepts to combine them, and outlines how to verify and simulate a system model defined using the combined graphical views of its con-
stituents, by considering the case study of an Isollete System.

Keywords: AADL · Simulink/Stateflow · Co-simulation ·
Code generation · Analysis

1 Introduction

Cyber-physical systems (CPSs), networked embedded systems (e.g., IoT, sensor networks), exploit computing units to monitor and control physical processes via wired or
wireless communication. CPSs are omnipresent, from high-speed train control systems,
power and control grids, automated plants and factories, transportations, to ground,
sea, air and space. Most CPSs are entrusted with mission- and safety-critical tasks.

J.-P. Talpin is partially supported by Nankai University.

P. Ribeiro and A. Sampaio (Eds.): UTP 2019, LNCS 11885, pp. 109–129, 2019.
https://doi.org/10.1007/978-3-030-31038-7_6

Therefore, the efficient and verified development of safe and reliable embedded systems is a priority mandated by many standards, yet a notoriously difficult and challenging domain.

As to standards, model-based design (MBD) has become a predominant development approach in the embedded system industry. In the MBD methodology, the development of a system starts with a model, based on which extensive analysis and verification are conducted, so that errors can be identified and corrected as early as possible, and ideally before the system is implemented or built. Subsequently, abstract system-level models are refined to semantically more concrete models and to source code, by model-transformation.

The merits of MBD hence include at least the following:

- Complexity becomes tractable and controllable, thanks to system level abstraction.
- Errors can be identified and corrected at the very early stages of system design.
- Correctness and reliability can be guaranteed by refinement.
- Developers can fully reuse existing components and/or systems, to improve development efficiency even more.

Unsurprisingly, available formalisms and environments for CPS design are numerous, e.g., hybrid automata [8], Hybrid CSP (HCSP) [22,44], dynamic differential logic [33], hybrid Event-B [10,11], Ptolemy [34], Metropolis [9], Crescendo [21], C2E2 [18], etc. in academia; Simulink/Stateflow [1,2], Modelica [39], SCADE [3], Labview, etc., in industry; UML, SysML [4], MARTE [38] and so on, for MBD. Because of the tight coupling of hardware, software, and physics in CPS design, one has to model a complex CPS from the perspectives of functionality (software), physicality (physical environment and hardware platform), and architecture uniformly, but unfortunately, most of existing modelling techniques do not support all of these three aspects well and uniformly.

For instance, the Architectural Analysis & Design Language (AADL) [20] is an architectural-centric model-based language developed by SAE International. It features strong capabilities to describe the architecture of a system due to the pragmatic (and practice-inspired) effectiveness of combining software and hardware component models. Meanwhile, it also supports the formal description of discrete behaviour using its BLESS Annex. Thanks to its succinct syntax, effective functionality and facilitated extensibility (by annexes i.e. plugins), AADL has been widely exploited in various embedded system domains, e.g., avionics, automotive. However, the core of the AADL only supports modelling of embedded system hardware structures and abstraction of its relevant discrete behaviour relevant to verification. It does not support the description of the continuous physical processes to be controlled by the embedded system and its combination with software, although some attempts have been made [6,7].

By contrast, Simulink [1] is the de facto standard toolbox that has demonstrated strong capabilities for model-based analysis and design of signal processing systems. It contains a large palette of functional blocks and supports their composition by continuous-time synchronous data-flow, as well as an intuitive graphical modelling language reminiscent of circuit diagrams. It is thus appealing to practitioners and engineers for whom it is designed. Moreover, Stateflow [2] is a toolbox adding facilities for modelling and simulating reactive systems by means of hierarchical statecharts, extending Simulink's scope to event-driven and hybrid forms of embedded control.

However, Simulink/Stateflow can hardly model system architectures and hardware platforms. To address this issue, we complement Simulink with AADL to provide a unified graphical modelling formalism to support all the three perspectives of CPS design uniformly. An overview of the combination is given in Fig. 1. For each CPS system to be modelled, it will be characterized from three different layers: architecture layer, software layer and physical layer. The modelling process is sketched as follows:

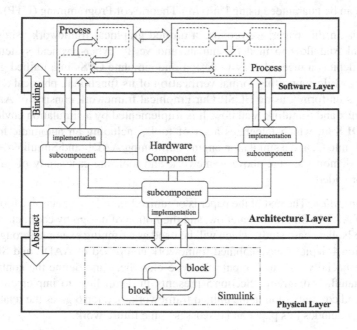

Fig. 1. An overview of the combination of AADL and Simulink

System architecture and hardware platform: arc given as AADL components in the architecture layer.

Software behaviour: is modelled either as AADL components or Simulink/Stateflow diagrams in the software layer.

Physical processes and its interaction with software: are modelled as Simulink/Stateflow diagrams in the physical layer.

Type classifier for Simulink/Stateflow diagrams: are generated as AADL components in the architecture layer. Given a Simulink/Stateflow diagram, a type classifier abstracts away the implementation details, and instead, defines the port declarations and the constraints for the behavior. The AADL abstract type classifier will be combined with the other AADL components to form the whole system in the architecture layer.

First, we translate Simulink/Stateflow diagrams into HCSP to obtain a formalisation of their port declarations [42, 46, 47]. Second, we use Daikon [19], or invariant generation [28] or, possibly, compositional proof tactics [30] to associate type classifiers with formal contracts.

Simulation of the whole graphical model, defined by the combination of AADL and Simulink views, amounts to coordinating code generated by both AADL and Simulink model simulators through effective port communications. Verification of the combined models is performed by translation to HCSP [14, 27, 41, 47]. Similar to the translation from Simulink/Stateflow to HCSP and the inverse [42, 46, 47], the correctness of the translation can be guaranteed using Unifying Theories of Programming (UTP) [15, 42].

Contribution. In this paper, we propose a unified graphical framework using AADL and Simulink/Stateflow to model, simulate and verify cyber-physical systems. This framework depicts a methodology to design and simulate CPSs in a unified graphical environment while supporting formal verification of its functional, physical and structural artifacts uniformly using HCSP. Our graphical framework consists of AADL, its BLESS Annex and Simulink/Stateflow. It is implemented by a simulation environment called **AADLSim**, which integrates a set of tools, including an automatic translator from AADL into C, and a simulation engine combining AADL and Simulink/Stateflow models. To demonstrate the above framework and tool, the case study of an Isolette System is provided.

Paper Organization. The rest of the paper is organized as follows. Section 2 provides an overview of AADL, Simulink/Stateflow, and the notion of design by contract. Section 3 presents the Isollete case study, which will be used as a running example throughout the paper. Section 4 depicts our combined framework composed of AADL and Simulink/Stateflow, especially how to compute the type classifiers and define the contracts for Simulink/Stateflow diagrams. Section 5 presents in detail how to implement the co-modelling and co-simulation in the unified framework. Section 6 gives the related work and Sect. 7 concludes this paper and discusses some future work.

2 Preliminaries

In this section, we first provide an overview of the AADL standard, by highlighting its structure and BLESS Annex, then introduce Simulink/Stateflow and its most relevant features. Finally, we briefly introduce the notion of design by contract.

2.1 AADL

AADL provides means to specify both the application software and the execution hardware of an embedded system, and supports textual, graphical and XML Metadata Interchange (XMI) specification formats. Components with *type* and *implementation* classifiers are instantiated and connected together to structure the system architecture. The AADL core language constructs are categorised into application software, execution platform and composite components. A *system* component represents a composite entity containing software, execution platform and system components.

Components and Connections. The execution platform category represents computation and communication resources including *processor*, *memory*, *bus* and *device* components. A processor component represents the hardware and software responsible for scheduling and executing task threads. Properties can be assigned to a processor component to specify scheduling policies, high-level operating system services and communication protocols. A memory component is used to represent storage entities for data and code. A device component can model a physical entity in the external environment: a plant or the software simulation of the plant. It can also be used as an interactive component like sensor or actuator. A bus component represents the physical connections among execution platform components.

The application software category includes *process*, *data*, *subprogram*, *thread*, and *thread group* components. A process component represents the protected address space, which is bound to a memory component. A data component can be used to abstract data type, local data or parameter of a subprogram. A subprogram models executable code that is called, with parameters, by thread and other subprograms. Thread is the only schedulable component with execution semantics to model system execution behavior. A thread represents sequential flow of the execution and the associated semantic automation describes life cycle of the thread.

A component type declaration defines interface elements and may contain *features*. Features comprise *data*, *event* and *event data* ports to transmit and receive data, control, and data/control respectively. Port communication is typed and directional. An *in* port receives data/control and an *out* port sends data/control while an *in out* port can send and receive data/control. Communication is realized through *connections* between ports, parameters and access to shared data.

BLESS Annex. The Behavior Language for Embedded System with Software (BLESS) is a standardised annex independent of the core AADL language. BLESS extends AADL with the ability of specifying behaviour of component interfaces, providing formal semantics for AADL behavioural descriptions and automatically generating verification conditions to be proven. BLESS models state machines using guards and actions to give precise specifications of discrete hardware/software behaviours. BLESS also introduces *assert* and *invariant* sections in AADL to specify assertions and predicates that behavioural models must satisfy.

We refer to AADL standard document AS5506-B [36] for further details.

2.2 Simulink/Stateflow

Simulink is an environment for model-based design of dynamical systems, and has become a de facto standard in the embedded systems industry. It provides an extensive library of pre-defined blocks for building and managing block diagrams, and also a rich set of fixed-step and variable-step solvers for simulating dynamical systems. It also provides features such as subsystems for building large systems in a hierarchical way. Stateflow is a toolbox adding facilities for modelling and simulating reactive systems by means of hierarchical statecharts. It extends Simulink scope to event-driven and hybrid forms of embedded control.

A Simulink model contains a set of blocks, subsystems, and wires, where blocks and subsystems cooperate by dataflow through the connecting wires. An elementary block receives input signals and computes output signals according to user-defined parameters altering its functionality. One typical parameter is sample time, which defines how frequently the computation is performed. Blocks are classified into two types: continuous blocks with sample time 0, and discrete blocks with positive sample time. For continuous blocks, the continuous state changes over time continuously, e.g. the position or the speed of a moving car. It is usually represented by an ordinary differential equation (ODE). Simulink provides an amount of ODE solvers for solving ODEs based on the numerical integration methods.

Stateflow offers the modelling capabilities of statecharts for reactive systems. It can be defined as Simulink blocks, fed with Simulink inputs and producing Simulink outputs. A stateflow diagram is composed of transitions, states and junctions. Each transition connects a source state to a destination state. It is labelled with $E[C]\{cAct\}/tAct$, where E is an event, C is the condition, $cAct$ the condition action, and $tAct$ the transition action. The event E triggers the transition to take place, provided that the condition C is true. As soon as C evaluates to true, the action $cAct$ will be executed immediately, while $tAct$ will be left pending and put in a queue first, and will be executed until a valid transition path is completed. A state is labelled by three optional types of actions: *entry action*, *during action*, and *exit action*.

Stateflow supports to construct *flow charts* using connective junctions and transitions, which can be used between states to specify decision logics to form transition networks. The Stateflow states can be composed to form hierarchical diagrams: *Or diagram*, for which the states are mutually exclusive and only one state becomes active at a time, and *And diagram*, for which the states are parallel and all of them become active simultaneously.

Being based on a large palette of individually simple function blocks and their composition by continuous-time synchronous dataflow as well as the modelling capabilities of statecharts for reactive systems, Simulink/Stateflow offers an intuitive graphical modelling language of CPSs for practicing engineers. Ordinary users can quickly build the model's framework by connecting the corresponding graphical modules and defining interfaces. Therefore, it is convenient and efficient to design and analyse the components using Simulink/Stateflow for co-simulation.

2.3 Design by Contract

Design by contract (DbC) is an engineering methodology whereby system designers should define semantically founded, precise and verifiable interface specifications for hardware and software components. These specifications extend the ordinary notion of abstract data type with logical properties describing the pre-conditions, post-conditions and invariants of a software function or of a hardware block.

The term design-by-contract is due to Bertrand Meyer in connection with the definition of the Eiffel programming language and his book Object-Oriented Software Construction [31]. It is rooted in Hoare logic, where the contract (A, G) of a program P naturally corresponds to the provable assertion $C \vdash \{A\}P\{G\}$ in some logical context

C. Contracts have been algebraically meta-theorized by Benveniste et al. [13], systematically applied to model-based design frameworks like BIP [12].

Recently, [30] extended the reach of design-by-contract to the case of modularly verifying hybrid system models by the introduction of contracts in a compositional design methodology for Differential dynamical Logic (ddL) [33]. In this context, and by contrast, the contract of a given model $\Gamma \vdash [\alpha]\phi$ consists of the evolution domain H of the specification α, as assumption, and differential invariant ϕ, as guarantee.

3   Isollete System: A Running Example

In this section, we introduce the *Isolette* System which we use as a running example. Isolette is an infant incubator described by the Federal Aviation Administration (FAA) in the Requirement Engineering Management Handbook (REMH) [26]. This example is concise but rich enough to contain both discrete control behaviour and continuous plants, as a classical hybrid system [7]. We will first introduce the system and then the design requirements.

3.1   Isollete System

The isollete example has been widely used to explain the detailed behaviour of AADL-based development and new annexes, such as the BLESS Annex and the Error Model Annex [17]. The isollete system is used to maintain the temperature of the isollete box, a physical environment, within a desired range that is beneficial to an infant.

Figure 3 depicts the AADL graphical model of the isollete system. The architecture of the system includes a processor, a bus, a sensor, an actuator, a controller, and a controlled process with internal threads. The software level defines the implementation of the controller, which obtains the temperature inside the box through the sensor, then computes an appropriate command to control the temperature through the actuator to switch on or off the heater combined with the isollete box. The physical layer defines the continuous behavior of the plant, i.e. the isollete box.

The continuous evolution of temperature depends on the current status of the actuator. If the heater is **on**, the temperature will increase, otherwise decrease. According to the specification in the Section A.5.1.3 of the REMH [26], when the isollete is properly switched on, the temperature of the heater will change at a rate of no more than $1\,°F$ *per minute*. Based on this specification, the temperature of the isollete box (denoted by c) and the temperature of the heater (denoted by q) are formally modelled by the ODEs (1) below.

$$\begin{cases} \dot{c} & = & -0.026 \cdot (c - q) \\ \dot{q} & = & 1 & \text{if heater is on} \\ \dot{q} & = & -1 & \text{if heater is off} \end{cases} \qquad (1)$$

The constant 0.026 stands for the thermal conductivity. When the controller commands the actuator to switch the heater on, the rise in temperature q will result in c going up. In this specification, we assume that the room temperature outside the box to be constant at $73\,°F$, although its variations could also be modelled.

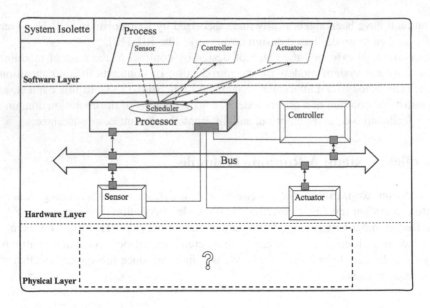

Fig. 2. AADL graphical model of isolette system

3.2 Requirements

Referring to environmental assumptions provided in the REMH, the following safety should be satisfied.

– **Safety:** The temperature inside the isollete box should be kept in between 97 °F and 100 °F, i.e., 97 °F $\leq c \leq$ 100 °F.

Moreover, considering the uncertainties from initial states, sensor errors, disturbance of dynamics, and numerical error caused by floating-point calculation, etc., it is required that:

– **Stability and Robustness:** The inside temperature c will be finally steered towards the valid range after some time.

At this point, it is obviously hard to specify this physical model using AADL and its annexes alone, notwithstanding its interaction with the digital controller, hence the question mark in Fig. 2 needs a complementary hybrid annex.

4   Combination of AADL and Simulink/Stateflow

The combination of AADL and Simulink/Stateflow aims at providing a unified graphical co-modelling formalism for CPSs, with which software, physical environment and execution hardware of a CPS can be modelled in a uniform framework. Section 4.1 presents a general explanation of the combined framework, Sects. 4.2 and 4.3 define the type classifiers for given Simulink/Stateflow diagrams, including the port declarations and contracts, respectively.

4.1   General Framework

As shown in Fig. 1, we describe the high-level architecture of the proposed unified graphical framework together with the connection among the three different physical, hardware and software layers. The architecture layer, described as AADL system composite components, specifies the types of hardware and software components, and (part of) their implementation (an abstraction of their actual implementation), as well as their composition. It usually consists of a central processor unit classifier with several subcomponent devices (like sensor, controller, and actuator etc.). Each of these classifiers has its own type and implementation. For software functionality and physical processes, the architecture layer usually needs their *abstractions*, i.e., the *type classifiers* of these software and physical components. The type classifier of a component declares the set of input and output ports, specifies the contract of its behaviour, that are accessible from outside. By contrast, the implementation classifier of a component binds its type classifier with a concrete implementation in the software and physical layers.

Our framework provides two methods to describe the type classifier of a given Simulink/Stateflow model. The first one is to derive a type classifier, which is satisfied by the Simulink/Stateflow diagram, directly from its behaviour; see Sect. 4.2; the other is to define a contract in the style of an assume/guarantee pair, and then prove the given Simulink/Stateflow diagram satisfying this contract; see Sect. 4.3.

In the software layer, software components are defined by their functionality, which can be done using either AADL or Simulink/Stateflow. In AADL, the functionality is defined by processes, and in each process, one or more threads may exist to describe specific controlling behaviours. The BLESS Annex can further be employed to specify the behavior of the system precisely. In order to establish a stable communication between different processes, a port declaration must be defined. The AADL implementation in this layer binds to the corresponding software and hardware components in the architecture layer.

In the physical layer, the continuous behaviour of physical processes is implemented as Simulink/Stateflow diagrams. In order to integrate the Simulink/Stateflow diagrams for implementing software or physical processes into the architectural layer, we need to define a type classifier for each Simulink/Stateflow diagram so that it can be assembled with other abstract components to form the architecture of the whole system at the architecture layer. We will explain the details of this process in the rest of this section.

Example 1. Now we can build a complete graphical model of the Isollete system with the combination as shown in Fig. 3, in which the Simulink/Stateflow diagram is given as Fig. 4.

Figure 4 implements the ODEs defined in (1). It receives the heat command from the actuator, depending on which the heater temperature q is implemented by an integrator block. The other integrator block computes the temperature c for the isollete box, which will be sent back to the sensor of the controller. This implementation will be abstracted as a AADL type classifier, to fill the definition of the isollete box in the physical layer in Fig. 3.

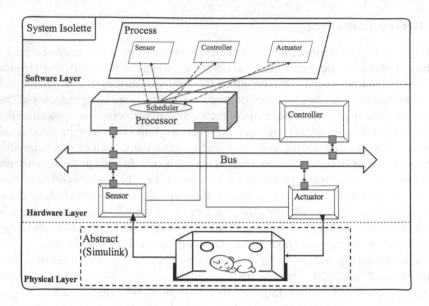

Fig. 3. AADL graphical model of isolette system

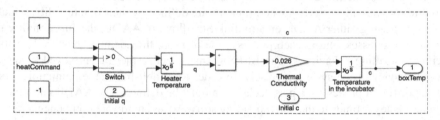

Fig. 4. Simulink model of isollete box

Simulation. To simulate the graphical model with the combination of AADL and Simulink, we propose a cross-layer co-simulation framework, in which the hardware platform, control software, and physical dynamics in the designed CPS can be taken into account uniformly. We will explain the details of such specification in Sect. 5.

Verification. To further verify a graphical model given by the AADL-Simulink combination, we translate it into HCSP, which is an extension of CSP introducing differential equations to model the continuous evolution of the plant and three types of interrupts to model the interaction between continuous and discrete behaviours [22,44]. The formal verification of HCSP can be done along the lines of our previous work [14,27,41,45]. Moreover, the correctness of the translation from Simulink/Stateflow to HCSP can be strictly proved using higher-order UTP [15], which extends the classic Unifying Theories of Programming (UTP) [23] to hybrid systems by introducing higher-order quantifications and differential relations. The technical details of this part will be reported in another paper.

4.2 Computing Type Classifier for Simulink/Stateflow Diagrams

As we explained above, when combining Simulink/Stateflow with AADL, we need to provide an abstraction for each Simulink/Stateflow diagram, i.e., its type classifier, so that it can be assembled with other components to form the whole system at the architecture layer, while the diagram itself will be used as the implementation classifier of the component. Normally, the type classifier of a component consists of two parts: *port declaration* and *constraints*.

The *port declaration* declares a set of ports used to input and output data between the component and other ones. However, Simulink diagrams can be hierarchical, and hence its external ports can sometimes not be extracted directly. For example, consider the triggered subsystems in a Simulink diagram, they do not have any input and output ports, but are triggered by events. Therefore, we need to analyse the whole system in detail in order to obtain all external ports, particular, event ports. Moreover, this often gets worse when Stateflow models are additionally considered.

To address this problem, we exploit the tool `Sim2HCSP`, a component in our toolkit MARS [14], which can translate a Simulink/Stateflow diagram into a formal HCSP process. By applying `Sim2HCSP`, all external ports of a Simulink/Stateflow diagram can now be translated, and exposed, by a set of channels in the corresponding HCSP model, which is stored in a separate file. In the case of the Simulink diagram in Fig. 4, we can for instance obtain the following port declaration:

$$\text{heatCommand}?q; \cdots ; \text{boxTemp}!c$$

from which the abstract type for `babybox` can be defined correspondingly:

```
abstract babybox
features
   heatCommand: in data port;
   boxTemp: out data port;
end babybox;
```

The reminder of the specification defines the contract of the component. It specifies the properties that should be satisfied by any execution of the component. In this paper, we adopt two approaches to generate the constraints for a given Simulink/Stateflow diagram. The first one uses Daikon [19]. The basic idea is to simulate the given Simulink/Stateflow diagram, and then run Daikon to generate a candidate invariant which is satisfied by all simulation runs. The more simulations are performed the more refined the generated invariant becomes. For example, considering the Simulink diagram in Fig. 4, by applying Daikon, we can obtain the following type classifier:

```
assert
  <<TIME: :(t >= 0.1)>>
  <<HEAT_T: :
     ((1.35107*10**15)*t-(1.35107*10**15)*q+9.862881*10**16=0)>>
  <<H_VAR: :
     ((2.111*10**13)*q-(2.111*10**13)*orig(q)+1.056*10**12=0)>>
  <<TEMP_VAR: :
     ((2.463*10**11)*c-(2.744*10**11)*orig(c)+2.055*10**12=0)>>
invariant
  <<TIME() and H_VAR() and H_VAR() and TEMP_VAR()>>
```

Alternatively, we can generate invariants directly from the Simulink/Stateflow diagram, or the translated HCSP process, by using techniques for invariant generation for hybrid systems, e.g., [28]. Consider again the Simulink diagram in Fig. 4. By using invariant generation, we can instead obtain the following type classifier:

```
assert
    <<L_LIMIT:  :  ((q-c)*e**(-0.026*t)+q*(c-97)<=0)>>
    <<H_LIMIT:  :  ((q-c)*e**(-0.026*t)+q*(100-c)>=0)>>
invariant
    <<L_LIMIT() and H_LIMIT()>>
```

The efficiency of the first approach is much higher, but the generated invariant (approximation) can only be linear. Moreover, it may not become an actual invariant, even by conducting enough runs to refine it. By contrast, the second approach can generate more expressive and semantically correct invariants, but the efficiency is normally very low. Improving the efficiency of invariant generation for hybrid systems is still a challenging problem.

4.3 Defining Type Classifier as Contracts

Our goal is to exploit the HCSP model provided by Sim2HCSP [14] to support modular, component-wise analysis and verification of system models combined from architectures described in AADL and hybrid systems in Simulink/Stateflow.

For HCSP models, our definition of contracts will naturally follow along the lines proposed by Lunel et al. for differential dynamic logic [29,30]. In that context, the contract of a specification α is defined by a pair (A, G) of properties. A, the assumption, is a formula defining the evolution domain of α and G, the guarantee, is a formula stating its differential invariant.

An alternative approach is to use the Hybrid Hoare Logic of HCSP [27,40]. In the HHL, the contract of an HCSP process P can be defined by the term $\{Pre\} P \{Post; HF\}$, where $Pre, Post$ represent the pre- and post-conditions of P using first-order logic and HF its history formula using the duration calculus. This not only allows to express properties upon start and finish but also real-time and continuous invariants on the execution of P, resulting in an undoubtedly more expressive framework, however probably challenging for proof automation.

In either approaches, it is hence tempting to investigate an adaptation of the composition theorem proposed in [30] to the HCSP framework, as it provides a methodology to automate the proof of a system contract, e.g. $(A_1 \wedge A_2, G_1 \wedge G_2)$, from the (possibly tedious) proofs that its components, e.g. $C_{1,2}$ satisfy the differential invariant $G_{1,2}$ in the evolution domains $A_{1,2}$, respectively.

This theorem is obtained using the parallel composition defined in [30, Def. 7], which amounts to decomposing the components $C_i \,\hat{=}\, \mathbf{disc}_i \cup \mathbf{cont}_i^*$ into discrete and continuous specifications $\mathbf{disc}_i$ and $\mathbf{cont}_i$, and recompose them as:

$$C_1 \otimes C_2 \,\hat{=}\, (\mathbf{disc}_1 \cup \mathbf{disc}_2 \cup (\mathbf{cont}_1, \mathbf{cont}_2)^*)$$

Assuming proof trees $\Gamma_i \vdash [C_i]G_i$, stating that the G_is are invariants of the components C_is in contexts Γ_i, for all $i = 1, 2$, and assuming non-interference of the definitions between the C_is nor with the guarantees of the G_js ($i \neq j$), [30, Th. 2] exhibits the derivation of a contract for the composed components: $\Gamma_{i=1,2} \vdash [\otimes_{i=1,2} C_i](\wedge_{i=1,2} G_i)$, yielding an automated proof tactic.

This model of compositional contracts can be employed to implement Sangiovanni-Vincentelli's "meet in the middle" design methodology [37] to mitigate software, hardware and physics constraints at system architecture level. In the case of the isolette, for instance, it can be used to verify the safety requirement of the isollette in nominal mode, Sect. 3.2 (i.e. after an initialization period) by cross-validating the differential invariant of the physical model with the (adequate) operations of its controller on the sensors and actuators, all four expressed by separate logical contracts. A use case of this method with KeymaeraX, concerning the well-known controlled water-tank problem, is given in Lunel's PhD Thesis [29].

5   Co-modelling and Co-simulation

This section details the implementation of the unified framework introduced in Subsect. 4.1 for designing and analyzing CPSs. The design flow of the framework is shown in Fig. 5. It consists of three stages: co-modelling, model translation, and co-simulation.

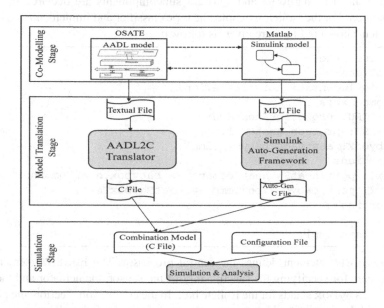

Fig. 5. Co-modelling and co-simulation of AADL and Simulink/Stateflow

In the co-modelling stage, designers can exploit the toolkit OSATE/AADL and Matlab/Simulink/Stateflow to model different parts of systems. Port definitions are required in each of the separate parts in order to establish the connection between AADL and Simulink/Stateflow models. Then, in the model translation stage, in order to integrate the two separate models and analyse them as a whole, we translate both the AADL and Simulink/Stateflow models into C code. For AADL, we developed a toolkit named *AADL2C Translator*, a novel code generation plugin to parse the AADL standard textual file and translate it into C code. Auto generation of C code from Matlab/Simulink/Stateflow models is done directly by using the Real Time Workshop (RTW) toolbox of Matlab. In the co-simulation stage, the translated C code from AADL and Simulink will be combined at first by performing integration and parameter configuration. After that, the generated model code will be compiled by a C compiler, with the co-simulation results produced. The co-simulation results provide a feedback for engineers to analyse and revise their original designs in the modelling stage. All three stages are introduced in detail in the subsequent subsections.

5.1   Co-modelling in AADL and Simulink/Stateflow

AADL Modelling. The OSATE platform provides two different approaches for engineers to build AADL models: graphical models and textual code. To exploit the internal mechanism of AADL, we choose the textual form to build our system. With the BLESS Annex, AADL is also able to specify the discrete behaviour of components. AADL adopts a top-down pattern to build a system: a *system* classifier is defined at the beginning, and then all its hardware and software subcomponents are declared in *system implementation*. For the Isollete example, the type classifier and implementation for the whole system shown in Fig. 3 are given as follows:

```
system isollete
end isollete;
system implementation isollete.impl
  subcomponents
    heatCPU: processor heatCPU;
    heatSW: process heatSW.impl;
    babybox: abstract babybox.impl;
  connections
    cnx1: port heatSW.heatCommand -> babybox.heatCommand;
    cnx2: port babybox.boxTemp -> heatSW.boxTemp;
  properties
    ......
end isollete.impl;
```

The `heatCPU` element defines the central processor. The `heatSW` element, the central process for specifying the functionality of the sensor, the actuator and the controller. The `babybox` stands for the isollete box. In the connections section, the ports of `heatSW` and the ports of `babybox` are connected, for transferring the heat command (representing the off or on status of the controlled variable) and the box temperature respectively. The properties section stipulates a binding relationship between the software and hardware subcomponents. We omit the details of it here. The behaviours of the sensor, actuator and the controller are implemented as threads in AADL.

In particular, the model of the controller is defined as follows:

```
thread controller
  features
    measuredTemp: in data port;
    diff: out data port;
end controller;
thread implementation controller.impl
  properties
    Dispatch_Protocol => Periodic;
    Priority => 10;
    Deadline => 20ms;
    Period => 20ms;
  annex BLESS {**
    invariant <<true>>
    variables: gain ;
    states s : initial complete final state;
    transition t : s -[on dispatch]-> s
    { gain := 10;
      if(measuredTemp > 100) diff := gain*(measuredTemp - 100);
      elsif(measuredTemp < 97) diff :=gain*(measuredTemp - 97);
      else diff :=0; end if; };
  **};
end controller.impl;
```

The controller receives the measured temperature of the isollete box from the sensor via the input port measuredTemp, and sends the difference between the temperature and the threshold to the actuator via output port diff. As defined in the implementation, the controller is executed periodically every 20 ms, with the deadline and priority defined; Its functionality is defined using the BLESS Annex. The local variable gain defines the gain coefficient for computing the difference, and the transition system of the controller includes one state and one transition, which computes the difference diff depending on the measured temperature. After receiving the value of diff, the actuator will decide whether to turn on or off the heat.

Simulink/Stateflow Modelling. We use Simulink/Stateflow to model the continuous behaviour of the CPS under design. For the Isollete box, we need to model the continuous behaviour defined by the ODE (1). The Simulink diagram has been given in Fig. 4.

Combination of Models. After building the models separately in AADL and Simulink/Stateflow, we combine them to form the whole system. Our approach is to define *abstract* components in AADL and connect each of them to the corresponding Simulink/Stateflow models. For each abstract component, the type classifier declares all the ports connecting AADL and Simulink/Stateflow models, and the constraints for the actual behaviour. The abstract AADL type classifier of the isollete box is given in Sect. 4.2.

5.2 Model Translation to C

Translating the AADL Model. The translation from AADL to C is the most crucial part in the unified framework. It uses a collection of mapping rules from AADL concepts to C implemented by the compiler *AADL2C Translator*. Figure 6 illustrates the model translation flow from AADL to C. A graphical model only describes the high-level architecture, while a textual model includes the details such as the functional behaviours.

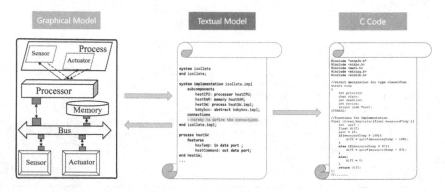

Fig. 6. An example illustrating the model translation flow from AADL to C

The compiler *AADL2C Translator* receives a source file as input and automatically generates the corresponding C code. According to the AADL grammar, a model is usually composed of several components, each with two parts: type declaration and implementation. The *AADL2C Translator* creates a `struct` object for each type declaration, e.g..system, process, thread, etc., and defines a collection of properties of the corresponding type classifiers. The implementation classifier is translated into individual sub-functions, associated with relevant type classifiers and specific names. Especially, for a thread implementation, two extra functions are specified for thread scheduling: *create_thread()* and *thread_scheduling()*. The *create_thread()* function adds a thread to a thread queue for dispatch, while *thread_scheduling()* is designed to execute threads according to the scheduling protocol specified in the AADL model, e.g., Rate-Monotonic Scheduling (RMS), Highest Priority First (HPF), etc.

In order to implement the port communications between different components efficiently, an additional Global Port Data Management (GPDM) unit is introduced in the target C code. The GPDM will store all of the output ports as global variables, with names of the form *componentType_componentName_outputPortName*. In each simulation cycle, the values of these variables will be updated once.

Translating Simulink/Stateflow Model. Matlab provides an automatic code generation tool that helps to translate Simulink/Stateflow models into C code. It greatly improves the quality and efficiency of development and simulation. To apply the code generation tool provided by Matlab, we need to set some configuration parameters, such as the model solver, the format of the generated code, etc.

Co-simulation requests a starting point of program execution so that we specify a main method as an interface to connect the C code generated from Simulink/Stateflow model with the C code interface generated from AADL model.

5.3 Co-simulation

Now all the different parts of the system, including hardware components, application softwares and physical processes modelled in AADL and Simulink/Stateflow, have been translated into C code separately. In the co-simulation stage, we need to integrated all the separate C code files by defining the communication between them. The communication of distributed parts is implemented through the GPDM block mentioned above, which can be regarded as a global data memory storing all the data interfaces (external in/out ports) information of each component, such as AADL thread components and Simulink/Stateflow models. Local variables inside components are not considered by the GPDM block.

After the C code files are integrated, the simulation of the whole system can be performed. At the beginning of the simulation, we need to set some configuration parameters, including the global simulation clock, the periodical simulation clock, the initial values of the system variables, and so on.

Simulation Results of Isollete. For the Isollete case study, we check whether it fulfils the requirements mentioned in Sect. 3.2 by simulation. We first translate the AADL and Simulink model of the Isollete to C code, then consider two cases by setting different initial values for the variables. In the first case, both the temperature inside isollete box and heat actuator are initially set as 73 °F, same as the general room temperature. In the second case, the initial temperature inside the isollete box is set as 115 °F (higher than the maximum safety temperature), and the temperature for the heat actuator is still set as 73 °F. For both of them, the simulation period is set to 0.1 s, and the simulation time is set to 300 s. Figures 7(a) and 7(b) show the simulation results for the two cases respectively, where the blue solid curve and yellow dashed curve represent the trajectories for continuous variables c (for box temperature) and q (for heat temperature) respectively. The simulation results show that, under the control of heat actuator, the temperature inside the isollete box will finally reach a stable state, within the safety range between 97 °F and 100 °F. The requirement defined in Sect. 3.2 is satisfied.

6 Related Work

AADL provides the notion of annex to support extensions to its core language. The key standardized annexes include the Behaviour Annex (BA) which extends AADL with the ability of defining component behaviour via state machines, and the BLESS Annex [25], which improves the state transition formalism by introducing assertions for supporting contract-based specifications. The simulation and analysis of AADL models

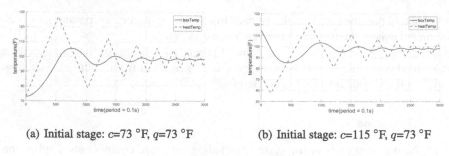

(a) Initial stage: c=73 °F, q=73 °F   (b) Initial stage: c=115 °F, q=73 °F

Fig. 7. Results of co-simulation from different initial stages

have also been explored a great deal. ADeS is a simulation tool that considers the environment in which the system evolves. AADL Inspector, produced by the Ellidiss company, is a powerful software that encompasses various features including schedulability analysis and dynamic simulation. There have also been some works on translation of AADL to other languages for analysis and simulation, e.g. AADL to BIP [16], AADL to Sync [24], AADL to Maude [32] and so on. However, most of them focus on the discrete-time behaviours.

There have been some works on the extension of AADL for hybrid systems. [43] models hybrid systems with AADL based on networks of timed automata, and uses the model checker UPPAAL for property analysis. [35] discusses a sublanguage extension to AADL to describe continuous behaviour, but it has difficulty in modelling complex continuous behaviour expressed with differential equations. In [5], a Hybrid Annex is presented, which is much more expressive in its ability to specify hybrid systems, yet it lacks relevant tools for further simulation and analysis of the hybrid models.

Compared with the above mentioned works, our proposed AADLSim framework coalesces AADL with Simulink for modelling of both discrete and continuous behaviours, and the flexible interaction between them. Moreover, it also provides extensive support for the analysis and simulation of the combined models through translating them into the same target language.

7   Conclusion and Future Work

In this paper, we propose an unified graphical co-modelling and co-simulation framework for the design of cyber-physical systems. This proposed framework combines AADL and Simulink/Stateflow with which the gap between discrete control and continuous plant can be filled. The combined models are translated into C code for further analysis by co-simulation. Throughout the paper, we clarify the main concepts of the framework, outline the specific co-simulation flow and the verification process of the combined models. An Isollete system case study is provided to illustrate the framework.

For future work, we will investigate the translation of the combination into HCSP, a formal modelling language encoding hybrid system dynamics by means of an extension of CSP. Formal verification of HCSP is supported by an interactive Hybrid Hoare Logic prover based on Isabelle/HOL. As a consequence, the combined AADL and

Simulink/Stateflow models can be verified. To make sure that the generated HCSP model is correct, the consistency between observational behaviours of the models at AADL and Simulink/Stateflow, and HCSP must be guaranteed in a rigorous way. This question, however, is known to be difficult, due to the inherent complexity of hybrid systems. To solve this problem, we consider to define the semantics of all the separate models in higher-order UTP [15], which extends the classic Unifying Theories of Programming (UTP) [23] to hybrid systems by introducing higher-order quantifications and differential relations. Moreover, we need to show that the domain of hybrid designs proposed in [15] together with the operations over hybrid designs forms a complete partial order (CPO), therefore can be used as a semantic domain.

References

1. Simulink User's Guide (2013). http://www.mathworks.com/help/pdf_doc/simulink/sl_using. pdf
2. Stateflow User's Guide (2013). http://www.mathworks.com/help/pdf_doc/stateflow/sf_ug. pdf
3. Esterel Technologies, SCADE suite (2018). http://www.esterel-technologies.com/products/ scade
4. SysML 1.6 Beta Specification (2019). http://www.omg.org/spec/SysML
5. Ahmad, E., Dong, Y., Larson, B., Lü, J., Tang, T., Zhan, N.: Behavior modeling and verification of movement authority scenario of Chinese train control system using AADL. Sci. China Inf. Sci. **58**(11), 1–20 (2015)
6. Ahmad, E., Dong, Y., Wang, S., Zhan, N., Zou, L.: Adding formal meanings to AADL with hybrid annex. In: Lanese, I., Madelaine, E. (eds.) FACS 2014. LNCS, vol. 8997, pp. 228–247. Springer, Cham (2015). https://doi.org/10.1007/978-3-319-15317-9_15
7. Ahmad, E., Larson, B.R., Barrett, S.C., Zhan, N., Dong, Y.: Hybrid annex: an AADL extension for continuous behavior and cyber-physical interaction modeling. In: ACM SIGAda Ada Letters, vol. 34, pp. 29–38 (2014)
8. Alur, R., Courcoubetis, C., Henzinger, T.A., Ho, P.-H.: Hybrid automata: an algorithmic approach to the specification and verification of hybrid systems. In: Grossman, R.L., Nerode, A., Ravn, A.P., Rischel, H. (eds.) HS 1991-1992. LNCS, vol. 736, pp. 209–229. Springer, Heidelberg (1993). https://doi.org/10.1007/3-540-57318-6_30
9. Balarin, F., Watanabe, Y., Hsieh, H., Lavagno, L., Passerone, C., Sangiovanni-Vincentelli, A.: Metropolis: an integrated electronic system design environment. Computer **36**(4), 45–52 (2003)
10. Banach, R., Butler, M., Qin, S., Verma, N., Zhu, H.: Core hybrid Event-B I: single hybrid Event-B machines. Sci. Comput. Program. **105**, 92–123 (2015)
11. Banach, R., Butler, M., Qin, S., Zhu, H.: Core hybrid Event-B II: multiple cooperating hybrid Event-B machines. Sci. Comput. Program. **139**, 1–35 (2016)
12. Basu, A., et al.: Rigorous component-based system design using the BIP framework. IEEE Softw. **28**(3), 41–48 (2011)
13. Benveniste, A., et al.: Contracts for system design. Found. Trends Electron. Des. Autom. **12**(2–3), 124–400 (2018)
14. Chen, M., et al.: MARS: a toolchain for modelling, analysis and verification of hybrid systems. In: Hinchey, M.G., Bowen, J.P., Olderog, E.-R. (eds.) Provably Correct Systems. NMSSE, pp. 39–58. Springer, Cham (2017). https://doi.org/10.1007/978-3-319-48628-4_3

15. Chen, M., Ravn, A.P., Wang, S., Yang, M., Zhan, N.: A two-way path between formal and informal design of embedded systems. In: Bowen, J.P., Zhu, H. (eds.) UTP 2016. LNCS, vol. 10134, pp. 65–92. Springer, Cham (2017). https://doi.org/10.1007/978-3-319-52228-9_4

16. Chkouri, M.Y., Robert, A., Bozga, M., Sifakis, J.: Translating AADL into BIP - application to the verification of real-time systems. In: Chaudron, M.R.V. (ed.) MODELS 2008. LNCS, vol. 5421, pp. 5–19. Springer, Heidelberg (2009). https://doi.org/10.1007/978-3-642-01648-6_2

17. Delange, J., Feiler, P.: Architecture fault modeling with the AADL error-model annex. In: 40th EUROMICRO Conference on Software Engineering and Advanced Applications, pp. 361–368. IEEE (2014)

18. Duggirala, P.S., Mitra, S., Viswanathan, M., Potok, M.: C2E2: a verification tool for stateflow models. In: Baier, C., Tinelli, C. (eds.) TACAS 2015. LNCS, vol. 9035, pp. 68 82. Springer, Heidelberg (2015). https://doi.org/10.1007/978-3-662-46681-0_5

19. Ernst, M.D., et al.: The Daikon system for dynamic detection of likely invariants. Sci. Comput. Program. **69**(1–3), 35–45 (2007)

20. Feiler, P.H., Gluch, D.P.: Model-Based Engineering with AADL: An Introduction to the SAE Architecture Analysis & Design Language. Addison-Wesley Professional, Boston (2012)

21. Fitzgerald, J., Larsen, P.G., Verhoef, M. (eds.): Collaborative Design for Embedded Systems: Co-modelling and Co-simulation. Springer, Heidelberg (2014). https://doi.org/10.1007/978-3-642-54118-6

22. He, J.: From CSP to hybrid systems. In: A Classical Mind, pp. 171–189. Prentice Hall International (UK) Ltd. (1994). Essays in Honour of C.A.R. Hoare

23. Hoare, C.A.R., He, J.: Unifying Theories of Programming. Prentice Hall, Upper Saddle River (1998)

24. Jahier, E., Halbwachs, N., Raymond, P., Nicollin, X., Lesens, D.: Virtual execution of AADL models via a translation into synchronous programs. In: EMSOFT 2007, pp. 134–143. ACM (2007)

25. Larson, B.R., Chalin, P., Hatcliff, J.: BLESS: formal specification and verification of behaviors for embedded systems with software. In: Brat, G., Rungta, N., Venet, A. (eds.) NFM 2013. LNCS, vol. 7871, pp. 276–290. Springer, Heidelberg (2013). https://doi.org/10.1007/978-3-642-38088-4_19

26. Lempia, D.L., Miller, S.P.: Requirements engineering management handbook. National Technical Information Service (NTIS) (2009)

27. Liu, J., et al.: A calculus for hybrid CSP. In: Ueda, K. (ed.) APLAS 2010. LNCS, vol. 6461, pp. 1–15. Springer, Heidelberg (2010). https://doi.org/10.1007/978-3-642-17164-2_1

28. Liu, J., Zhan, N., Zhao, H.: Computing semi-algebraic invariants for polynomial dynamical systems. In: Proceedings of the Ninth ACM International Conference on Embedded Software, pp. 97–106. ACM (2011)

29. Lunel, S.: Parallelism and modular proof in differential dynamic logic. (Parallélisme et preuve modulaire en logique dynamique différentielle). Ph.D. thesis, University of Rennes 1, France (2019)

30. Lunel, S., Boyer, B., Talpin, J.: Compositional proofs in differential dynamic logic dL. In: ACSD 2017, pp. 19–28 (2017)

31. Meyer, B.: Object-Oriented Software Construction, 2nd edn. Prentice-Hall Inc., Upper Saddle River (1997)

32. Ölveczky, P.C., Boronat, A., Meseguer, J.: Formal semantics and analysis of behavioral AADL models in real-time maude. In: Hatcliff, J., Zucca, E. (eds.) FMOODS/FORTE - 2010. LNCS, vol. 6117, pp. 47–62. Springer, Heidelberg (2010). https://doi.org/10.1007/978-3-642-13464-7_5

33. Platzer, A.: Logical Foundations of Cyber-Physical Systems. Springer, Cham (2018). https://doi.org/10.1007/978-3-319-63588-0

34. Ptolemaeus, C. (ed.): System Design, Modeling, and Simulation using Ptolemy II. Ptolemy.org (2014)
35. Qian, Y., Liu, J., Chen, X.: Hybrid AADL: a sublanguage extension to AADL. In: Internetware 2013. ACM (2013)
36. SAE International Standards: Aarchitecture analysis & design language (AADL), revision B (2012)
37. Sangiovanni-Vincentelli, A.: Quo vadis, SDL: reasoning about trends and challenges of system-level design. Proc. IEEE **95**(3), 467–506 (2007)
38. Selic, B., Gerard, S.: Modeling and Analysis or Real-Time and Embedded Systems with UML and MARTE: Developing Cyber-Physical Systems. The MK/OMG Press, Boston (2013)
39. Tiller, M.: Introduction to Physical Modeling with Modelica. The Springer International Series in Engineering and Computer Science. Springer, Boston (2001). https://doi.org/10.1007/978-1-4615-1561-6
40. Wang, S., Zhan, N., Guelev, D.: An assume/guarantee based compositional calculus for hybrid CSP. In: Agrawal, M., Cooper, S.B., Li, A. (eds.) TAMC 2012. LNCS, vol. 7287, pp. 72–83. Springer, Heidelberg (2012). https://doi.org/10.1007/978-3-642-29952-0_13
41. Wang, S., Zhan, N., Zou, L.: An improved HHL prover: an interactive theorem prover for hybrid systems. In: Butler, M., Conchon, S., Zaïdi, F. (eds.) ICFEM 2015. LNCS, vol. 9407, pp. 382–399. Springer, Cham (2015). https://doi.org/10.1007/978-3-319-25423-4_25
42. Zhan, N., Wang, S., Zhao, H.: Formal Verification of Simulink/Stateflow Diagrams. Springer, Cham (2017). https://doi.org/10.1007/978-3-319-47016-0
43. Yu, Z., Yunwei, D., Fan, Z., Yunfeng, Z.: Research on modeling and analysis of CPS. In: Calero, J.M.A., Yang, L.T., Mármol, F.G., García Villalba, L.J., Li, A.X., Wang, Y. (eds.) ATC 2011. LNCS, vol. 6906, pp. 92–105. Springer, Heidelberg (2011). https://doi.org/10.1007/978-3-642-23496-5_7
44. Chaochen, Z., Ji, W., Ravn, A.P.: A formal description of hybrid systems. In: Alur, R., Henzinger, T.A., Sontag, E.D. (eds.) HS 1995. LNCS, vol. 1066, pp. 511–530. Springer, Heidelberg (1996). https://doi.org/10.1007/BFb0020972
45. Zou, L., et al.: Verifying Chinese train control system under a combined Scenario by theorem proving. In: Cohen, E., Rybalchenko, A. (eds.) VSTTE 2013. LNCS, vol. 8164, pp. 262–280. Springer, Heidelberg (2014). https://doi.org/10.1007/978-3-642-54108-7_14
46. Zou, L., Zhan, N., Wang, S., Fränzle, M.: Formal verification of simulink/stateflow diagrams. In: Finkbeiner, B., Pu, G., Zhang, L. (eds.) ATVA 2015. LNCS, vol. 9364, pp. 464–481. Springer, Cham (2015). https://doi.org/10.1007/978-3-319-24953-7_33
47. Zou, L., Zhan, N., Wang, S., Fränzle, M., Qin, S.: Verifying Simulink diagrams via a hybrid Hoare logic prover. In: EMSOFT 2013, pp. 1–9. IEEE (2013)

Hybrid Relations in Isabelle/UTP

Simon Foster[✉][iD]

University of York, York, UK
simon.foster@york.ac.uk

Abstract. We describe our UTP theory of hybrid relations, which extends the relational calculus with continuous variables and differential equations. This enables the use of UTP in modelling and verification of hybrid systems, supported by our mechanisation in Isabelle/UTP. The hybrid relational calculus is built upon the same foundation as the UTP's theory of reactive processes, which is accomplished through a generalised trace algebra and a model of piecewise-continuous functions. From this foundation, we give semantics to hybrid programs, including ordinary differential equations and preemption, and show how the theory can be used to reason about sequential hybrid systems.

1 Introduction

Cyber-Physical Systems (CPSs) use computation to monitor and control real-world phenomena, employing sensors and actuators. Autonomous mobile robots, for example, implement their goals by sensing the environment, updating an internal model of the real-world, using the model to plan and make decisions, and finally actuating. A common way of modelling, simulating, and verifying CPSs is with the use of a hybrid dynamical systems modelling language, such as Simulink[1], Modelica[2], hybrid programs[3] [1], and Hybrid CSP [2,3] (HCSP). Here, a system model is decomposed into two parts: (1) a digital controller, which is described used traditional programming constructs; and (2) a continuously evolving environment, which is described using differential equations.

Languages like Simulink and Modelica are used commercially for developing CPSs, since they are largely diagrammatic in nature and can be used to produce executable code. Typically, however, such tools support only simulation and testing, which limit their effectiveness for verification. On the other hand, tools like KeYmaera X [4] and HHL Prover [5] support rigorous formal verification, for differential dynamic logic [1] ($d\mathcal{L}$) and HCSP [2,3], respectively, that can prove properties of the entire state space symbolically. However, the latter tools are hard to apply for non-academics, and there is need for greater integration with the commercial tools [6]. A precondition of this is that there are

---

[1] Simulink: https://uk.mathworks.com/products/simulink.html.
[2] Modelica Language: https://www.modelica.org/modelicalanguage.
[3] The modelling notation of differential dynamic logic ($d\mathcal{L}$).

© Springer Nature Switzerland AG 2019
P. Ribeiro and A. Sampaio (Eds.): UTP 2019, LNCS 11885, pp. 130–153, 2019.
https://doi.org/10.1007/978-3-030-31038-7_7

unified semantic foundations for hybrid systems that acknowledge both similarities and differences between the languages, and integrated mechanised reasoning to support comprehensive automated formal verification.

The goal of this paper is to make first steps towards this foundation with a mechanised UTP theory for hybrid systems. UTP [7] is concerned with establishing formal links between languages based on heterogeneous computational paradigms, and therefore it is wholly appropriate to apply it to study of hybrid computational models. Our contributions are: (1) a UTP theory that incorporates a piecewise continuous timed trace model, building on our previous theory of generalised reactive relations [8]; (2) denotational semantics for a simple imperative language for hybrid programs, inspired by $d\mathcal{L}$ and HCSP; and (3) mechanised reasoning support in our UTP theorem prover, Isabelle/UTP [9–12]. Our hybrid theory represents a substantial overhaul of our previous results [13,14] by unifying it with our generalised UTP theory of reactive processes [8].

Most theorems and definitions in the paper are accompanied by a small Isabelle icon (🌀). In the electronic version, each icon is hyperlinked to the corresponding mechanised artefact in our Isabelle/UTP GitHub repository[4]. This, we hope, will convince the reader of the level of rigour employed in this work.

Our paper is structured as follows. Section 2 gives an overview of our hybrid program notation. Section 3 gives an overview of Isabelle/UTP, and how it is used to model programs. Section 4 presents our foundational UTP theory of generalised reactive processes. Section 5 describes a model for piecewise continuous timed traces, which is the basis for modelling continuous state spaces and variables. Section 6 gives a comprehensive exposition of the UTP theory of hybrid relations. Section 7 illustrates a small verification example in Isabelle/UTP. Section 8 concludes and discusses our results.

2   Hybrid Systems and Programs

In this section, we briefly introduce the key concepts of hybrid systems and programs, to the set the technical work that follows in context.

Hybrid systems exhibit both continuous flows and discrete jumps in the values of their variables [3,15]. Typically, a hybrid system evolves according to a differential equation, until some condition is satisfied, at which point a discrete jump occurs. For example, in the classic bouncing ball example, the ball is dropped and falls until it impacts the floor. During flight, the height above the ground, h, and velocity, v, change continuously. However, once the ball impacts the floor, the velocity is instantaneously inverted and it begins to travel upwards.

In this paper, we model such systems with a form of hybrid program:

Definition 2.1 (Hybrid Programs).

$$\mathcal{H} ::= ?b \mid x := v \mid \langle \dot{x}(t) \bullet f(t,x) \rangle \mid \mathcal{H} \, ; \mathcal{H} \mid \mathcal{H} \sqcap \mathcal{H} \mid \mathcal{H}^* \mid \mathcal{H} \bigtriangleup \langle b \mid c \rangle \mid \cdots$$

---

[4] Isabelle/UTP repository: https://github.com/isabelle-utp/utp-main.

where x is a name, t is a time variable, b and c are predicates, v is an expression.

As is common in UTP, the syntax can be further extended, which is the reason for the ellipsis. The simple language contains a mixture of constructs adapted from d$\mathcal{L}$ hybrid programs and HCSP. As usual, programs can be composed sequentially ($P \; ; \; Q$) and nondeterministically ($P \sqcap Q$). As in d$\mathcal{L}$, we can also define tests, $?b$, which execute when assumption b is satisfied, and assignment $x := v$. Hybrid programs can be iterative, which is expressed by the Kleene star P^*.

We characterise ordinary differential equations (ODEs), $\langle \, \dot{x}(t) \bullet f(t, x) \, \rangle$, which express that the derivative of the variable vector x is given by f. Its behaviour is to evolve the variables, without terminating, according to a solution $x(t)$, such that $\dot{x}(t) = f(t, x(t))$. For example, we can model a real-time clock by creating a distinguished continuous variable called *time*, such that $\dot{time}(t) = 1$.

Finally, $P \, \triangle \, \langle b \, | \, c \rangle$ allows preemption of a continuous evolution. Evolution of P may continue whilst b, a condition on the continuous variables, is invariant, and can terminate once c becomes true. The reason for having both b and c is to allow nondeterminism around when an evolution is preempted. Such nondeterminism exists in languages like Modelica, where discrete jumps are implemented using "zero-crossing detection", such that a function goes from positive to negative or vice-versa. This is subject to numerical imprecision, and thus the point at which the event occurs is effectively nondeterministic.

To illustrate hybrid programs, we formalise the bouncing ball example below:

Example 2.2 (Bouncing Ball as a Hybrid Program).

$$\text{BBall} \triangleq h := 2.0 \; ; \; v := 0 \; ; \; \begin{pmatrix} \langle \, (\dot{h}(t), \dot{v}(t)) \bullet (v, -9.81) \, \rangle \\ \triangle \, \langle h \geq 0 \, | \, v \leq 0 \wedge h < \epsilon \rangle \; ; \\ v := -0.8 \cdot v \end{pmatrix}^*$$

Initially, the height of the ball is 2 m, and the velocity is 0. Then, the main body of the system begins, by first evolving h and v according to a system of ODEs. The ODEs state that the derivative of h is v, and the derivative of v is -9.81, the standard gravity constant. Evolution continues whilst $h \geq 0$: the ball is above the ground, which gives a bound on the evolution. Once h falls below a constant $\epsilon > 0$, a small number which characterises numerical imprecision, and assuming $v \leq 0$ (flight is downwards), then the evolution can terminate. At this point, v is discontinuously inverted and a damping factor of .8 is applied. Through the Kleene star, the system is permitted to iterate zero or more times.

In the remainder of this paper, we show how such hybrid programs can be mechanically supported with our UTP theory of hybrid relations.

3   Isabelle/UTP

Isabelle/UTP [10,11] is a mechanisation of UTP in Isabelle/HOL, along with the main results from the UTP book [7] and related publications [16,17]. It provides an implementation of the alphabetised relational calculus, a model of imperative programs, a large library of algebraic laws, and several automated proof tactics. It mechanically supports the following activities: (1) development of UTP theories for languages of various computational paradigms; (2) construction of denotational semantics for said languages; and (3) creation of proof strategies to support automated verification tools using UTP theories. Aside from UTP, our mechanisation also draws heavily on the work of Back and von Wright [18].

For the imperative program model, Isabelle/UTP supports several calculi, including Hoare logic, weakest (liberal) precondition, and structural operational semantics. These axiomatic semantics are defined denotationally, and the associated laws proved as theorems. Linking theorems can also show correspondences between different semantic presentations. For example, it is well known that

$$\{\,b\,\}\,P\,\{\,c\,\} = (b \Rightarrow P\ \textbf{wlp}\ c)$$

is a theorem of Hoare calculus and weakest liberal preconditions [7,19]. Such a result can be harnessed to recast a verification theorem into a different form, which is potentially easier to prove. For each verification calculus, Isabelle/UTP also provides proof tactics, including deductive reasoning for Hoare logic, and equational reduction for **wlp**. The output of these tactics is a set of verification conditions (VCs), to which Isabelle's automated proof strategies can be applied.

A main objective in implementing Isabelle/UTP has been to harness the power of Isabelle's automated reasoning. Consequently, Isabelle/UTP follows in the tradition of shallow embeddings [20–22] in reusing as much as possible of the Isabelle technical infrastructure, such as its type system, parser, term language, and meta-logic, in defining the relational calculus. However, this objective must be reconciled with the need to provide a sufficiently expressive relation model to allow expression of UTP theories and the associated laws. The crucial artifact to get right here is the mechanisation of UTP variables and alphabets.

In Isabelle/UTP, state spaces are modelled as Isabelle types, and programs are parametric in their state space. Assuming a suitable state space Σ, a relational program is effectively modelled as a subset of $\mathbb{P}(\Sigma \times \Sigma)$, and an expression of type τ can be modelled as functions $\Sigma \to \tau$. This is consistent with most other works on verification using Isabelle/HOL [20,23], and allows us to obtain the UTP relational operators easily, such as disjunction $(P \lor Q)$, relational composition $(P\ ;\ Q)$, tests $(?b)$, and refinement $(P \sqsubseteq Q)$. However, this does not in itself provide us with a variable model. Rather than modelling these syntactically using names, we treat them as algebraic objects called lenses [11,18,24]:

Definition 3.1. *A lens is a quadruple* $\langle V \,|\, S \,|\, get : S \to V \,|\, put : S \to V \to S \rangle$, *where* V *and* S *are non-empty sets called the view and source, respectively, and* **get** *and* **put** *are total functions, such that the following equations hold:*

$$get\,(put\,s\,v) = v \qquad put\,(put\,s\,v')\,v = put\,s\,v \qquad put\,s\,(get\,s) = s$$

We write $V \Longrightarrow S$ *to denote the type of lenses with source type* S *and view type* V, *and subscript* **get** *and* **put** *with the name of a particular lens.*

Each variable $x : \tau$ in UTP is modelled using a lens $\tau \Longrightarrow \Sigma$, for some suitable state space type, and each get_x/put_x pair is used to query and update its value. The main advantage of this algebraic encoding is that we obtain an abstract representation that unifies several state space representations. We also note that a very similar concept to lenses exists in Back's refinement calculus [18, Chapter 5], which substantially predates the work on lenses [24].

With lenses, expressions can modelled as functions on the Σ; for example:

$$[\![x > (y + z)/2]\!] = (\lambda\, s : \Sigma \bullet get_x\, s > (get_y\, s + get_z\, s)/2)$$

In this way, each operator at the expression level corresponds to a point-wise lifting of the corresponding operator through the state s at the function level.

With lenses, we can also generically characterise several variable properties:

1. independence, $x \bowtie y$—x and y refer to disjoint views of Σ;
2. inclusion partial order, $x \preceq y$—the view of y contains the view of x;
3. equivalence, $x \approx y$—the lenses x and y refer to identical views.

All of these properties reduce to properties of the corresponding get and put functions; for example independence is essentially commutativity of put_x and put_y. They allow us to effectively characterise meta-logic properties of variables, which are normally characteristic of a deep embedding.

Variable updates are described using substitutions $\sigma : \Sigma \to \Sigma$, which are total functions on the state space, and allow us to describe assignments, variables contexts, and substitutions. The most basic substitution is the identity function, id, which effectively maps every variable to its present value. A substitution can be updated using the operator $\sigma(x \mapsto e)$, which associates x with an expression e over Σ. Then, we use the notation $[x_1 \mapsto e_1, \cdots, x_n \mapsto e_n]$ as a shorthand for $id(x_1 \mapsto e_1 \cdots x_n \mapsto e_n)$, which is a simultaneous substitution for n variables. Substitution update obeys several algebraic laws:

Theorem 3.2. *If x and y are lenses, then the following identities hold:*

$$\sigma(x \mapsto x) = \sigma \tag{3.2.1}$$

$$\sigma(x \mapsto e, y \mapsto f) = \sigma(y \mapsto f, x \mapsto e) \qquad \text{if } x \bowtie y \tag{3.2.2}$$

$$\sigma(x \mapsto e, y \mapsto f) = \sigma(y \mapsto f) \qquad \text{if } x \preceq y \tag{3.2.3}$$

(3.2.1) shows that a trivial update is ineffectual. (3.2.2) shows that two maplets may be commuted if the lenses are independent. (3.2.3) shows that a maplet for x is overridden by one assigning y when $x \preceq y$, and thus also when $x \approx y$.

Substitutions can be applied to expressions using $\sigma \dagger e$, which replaces all the variables in e with those assigned in σ. This is similar to syntactic substitution, with $e[v/x] = [x \mapsto v] \dagger e$, and obeys similar laws, but it is a semantic operator that composes σ with e (both are functions).

We also use substitutions to construct assignments, using Back's generalised operator [18]: $\langle \sigma \rangle$. This operator recasts the function σ as a relation. A singleton assignment, $x := v$, can be denoted using $\langle x \mapsto v \rangle$, and a simultaneous assignment by $\langle x_1 \mapsto v_1, x_2 \mapsto v_2, \cdots \rangle$. We can prove several familiar assignment laws:

Theorem 3.3 (Assignment Laws).

$$\langle \sigma \rangle \,;\, \langle \rho \rangle \;=\; \langle \rho \circ \sigma \rangle \tag{3.3.1}$$

$$x := x \;=\; \langle \textit{id} \rangle \tag{3.3.2}$$

$$x := e \,;\, y := f \;=\; y := f \,;\, x := e \qquad x \bowtie y, x \,\sharp\, f, y \,\sharp\, e \tag{3.3.3}$$

$$x := e \,;\, x := f \;=\; x := f[e/x] \tag{3.3.4}$$

The first law is a homomorphism law for assignments. The other laws are corollaries of it and the laws of Theorem 3.2. The third law, showing that assignments commute, requires an extra side condition that f does not mention x, and e does not mention y. These are both formulated using a semantic operator called unrestriction, $x \,\sharp\, f$, which means that f does not depend on the state space region characterised by x for its valuation, and is denoted using the lens operators [10].

Thus we have demonstrated the ubiquity of lenses in capturing the UTP relational calculus. In the next section we describe of theory of generalised reactive processes that is the foundation of the hybrid relational calculus. Later, we show how lenses are used to characterise continuous variables.

4   Trace Algebra and Generalised Reactive Relations

In this section, we describe our theory of generalised reactive relations. This UTP theory provides the foundation for our theory of hybrid relations using an abstract trace model. This, in particular, can be instantiated with piecewise continuous functions, which are often used to semantically capture the behaviour of hybrid systems [3,15].

The UTP theory of reactive processes [7,16] provides a generic foundation for trace-based reactive languages. Originally the trace model was fixed to discrete sequences, to support the semantic models of CSP and ACP [7]. In previous work [8], we generalised this theory to characterise traces abstractly with a *trace algebra*. We characterise traces with a set $\mathcal{T}$ and two operators: concatenation $\frown : \mathcal{T} \to \mathcal{T} \to \mathcal{T}$, and the empty trace $\varepsilon : \mathcal{T}$, which obey the following axioms [8].

Definition 4.1. *A* trace algebra $(\mathcal{T}, \frown, \varepsilon)$ *satisfies the following axioms:*   🐾

$$x \frown (y \frown z) = (x \frown y) \frown z \quad \text{(TA1)}$$

$$\varepsilon \frown x = x \frown \varepsilon = x \qquad\qquad \text{(TA2)}$$

$$x \frown y = x \frown z \;\Rightarrow\; y = z \quad \text{(TA3)}$$

$$x \frown z = y \frown z \;\Rightarrow\; x = y \quad \text{(TA4)}$$

$$x \frown y = \varepsilon \;\Rightarrow\; x = \varepsilon \quad \text{(TA5)}$$

TA5 ensures that every trace is positive $(x \geq 0)$; its lefthand dual is a theorem of these axioms. An example model is formed by finite sequences, $\langle a, b, \cdots, z \rangle$, that is $(\text{seq } A, \frown, \langle \rangle)$ forms a trace algebra, where $\frown$ is concatenation. Using the trace algebra operators, we can define trace prefix $(x \leq y)$, which partially orders traces, and trace difference $(x - y)$, which removes a prefix y from x [8].

From these algebraic foundations, we reconstruct the complete UTP theory of reactive processes [7,16], including its healthiness conditions and associated laws, in particular those for sequential and parallel composition [8]. For our version of the theory, the alphabet includes the following observational variables:

1. $ok, ok' : \mathbb{B}$ – to indicate whether there is divergence;
2. $wait, wait' : \mathbb{B}$ – to indicate whether a process is intermediate;
3. $tr, tr' : \mathcal{T}$ – to represent the trace, using a suitable trace algebra;
4. $st, st' : \Sigma$ – to represent the state, for some non-empty state space Σ.

We then define the following reactive healthiness conditions [7,16]:

Definition 4.2 (Reactive Relations Healthiness Conditions).

$$R1(P) \triangleq P \wedge tr \leq tr'$$

$$R2(P) \triangleq P[\langle\rangle, tr' - tr/tr, tr'] \lhd tr \leq tr' \rhd P$$

$$RR(P) \triangleq (\exists(ok, ok', wait, wait') \bullet R1(R2(P)))$$

$$RC(P) \triangleq R1(RR(P) ; tr' \leq tr)$$

$$tt \triangleq (tr' - tr)$$

The main healthiness conditions are RR, which describes reactive relations, and RC, which describes a subset of RR called reactive conditions. For our purposes, a reactive relation is, intuitively, a relation that refers to the initial and final values of state variables (x and x', where $x \preceq st$), and a special variable tt, that denotes a trace contribution. Technically, tt is an expression that denotes the difference $tr' - tr$, whose well-formedness is ensured by the commuting reactive healthiness conditions $R1$ and $R2$. This is reflected by the following theorem:

Theorem 4.3. *If P is RR healthy then* $P = (\exists\, t \bullet P[\varepsilon, t/tr, tr'] \wedge tr' = tr \,\widehat{}\, t)$

Any observation of a reactive relation P characterises a trace extension t which can be observed using tt. Reactive relations are closed under most relational operators; the exceptions are the universal relation (**true**), complement ($\neg$), and implication ($\Rightarrow$). These all require imposition of $R1$, and so we recast them as **true**$_r$, $\neg_r$, and $\Rightarrow_r$, respectively. Reactive relations form several algebras, including (1) a Boolean algebra [25], (2) a complete lattice [25], and (3) a Kleene algebra [26], which allows us to reason about iterative reactive programs (P^*).

The generalised assignment operator, $\langle\sigma\rangle$, is not in general healthy as it permits assignment of any variable, including ok, $wait$, and tr, which can violate RR. Consequently, we recast the operator $\langle\sigma\rangle_r$, where $\sigma : \Sigma \to \Sigma$ operates on the program state in st only. It obeys analogous laws to those in Theorem 3.3.

The second main healthiness condition in Definition 4.2, RC, characterises reactive conditions. A reactive condition is a reactive relation which (1) does not refer the final value of the state variables (st'), and (2) characterises a set of traces that is prefix closed. Reactive conditions are analogous to relational conditions, which refer to the initial state only, but can refer to both tr and tr', provided that

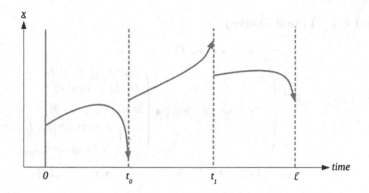

Fig. 1. Piecewise continuous timed traces

tt is prefix closed. For example, if we apply $\textbf{\textit{RC}}$ to the non prefix-closed relation $tr' = tr \frown \langle a \rangle$, then we obtain $\textbf{\textit{R1}}(tr' \leq tr \frown \langle a \rangle) = \textbf{\textit{R1}}(tt \in \{\langle \rangle, \langle a \rangle\})$, which is prefix closed. The intuition is that when a reactive condition is satisfied, it should also be satisfied by any prefix of the trace. Reactive conditions are particularly useful to characterise assumptions in our theory of reactive contracts [25], which extends reactive relations with assume/guarantee reasoning (see Sect. 6.5).

We have outlined our theory of reactive relations. In the next section we construct a trace algebra model that allows us to specialise to hybrid relations.

5   Continuous State and Timed Traces

In this section we describe how continuous state is modelled using a timed trace model, which characterises piecewise continuous trajectories [15]. Our model refines our previous work [8] by requiring that each continuous segment also converges, ensuring that its final value can be obtained. This requirement is always satisfied by, for example, linear ODEs. The state space (Σ) of a hybrid system consists of discrete variables, which exhibit only jumps at certain instants, and continuous variables, which change constantly with respect to time. Consequently, we subdivide the state space Σ into a discrete state space (Σ_d) and a continuous state space (Σ_c), both of which are non-empty, and then $\Sigma \triangleq \Sigma_d \times \Sigma_c$.

We require that Σ_c, minimally, forms a topological (Hausdorff) space, so that we can describe limits of a function over Σ_c. A special case is when $\Sigma_c = \mathbb{R}^n$, for some $n : \mathbb{N}$, that is a Euclidean state space. We impose no additional constraints on Σ_d which characterises variables that only exhibit discontinuous changes.

We now define our model of timed traces, which refines our previous model [8] by adding a convergence requirement:

Definition 5.1 (Timed Traces).

$$\mathbb{TT}_{\Sigma_c} \triangleq \left\{ f : \mathbb{R}_{\geq 0} \nrightarrow \Sigma_c \; \middle| \; \begin{array}{l} \exists\, t : \mathbb{R}_{\geq 0} \bullet \operatorname{dom}(f) = [0, t) \; \wedge \\ \left(t > 0 \Rightarrow \exists\, I : \mathbb{R}_{osq} \bullet \left(\begin{array}{l} \operatorname{ran}(I) \subseteq [0, t] \; \wedge \\ \{0, t\} \subseteq \operatorname{ran}(I) \; \wedge \\ \forall\, n \in [0, \#I - 2] \\ \quad \bullet \left(\begin{array}{l} f \ \textit{cont-on}\, [I_n, I_{n+1}) \\ \wedge f \ \textit{has-limit}\, I_{n+1} \end{array} \right) \end{array} \right) \right) \end{array} \right\}$$

where
$$\mathbb{R}_{osq} \triangleq \{ x : \operatorname{seq}\mathbb{R} \mid \forall\, n < \#x - 1 \bullet x_n < x_{n+1} \}$$

$$f \ \textit{cont-on}\, A \triangleq \forall\, t \in A \bullet \lim_{x \downarrow t} f(x) = f(t)$$

$$f \ \textit{has-limit}\, k \triangleq (\exists\, l : \mathbb{R} \bullet \lim_{x \uparrow k} f(x) = l)$$

A timed trace is a partial function from positive real numbers ($\mathbb{R}_{\geq 0}$) to the continuous state space, Σ_c, that satisfies certain constraints. Firstly, we require that the domain is a right open interval from 0 to some positive real t. Secondly, if t is non-zero, we require that the function is composed of a sequence of continuous segments, each of which converges to a limit. The intuition is sketched in Fig. 1 for a state space with a single variable x. There are three continuous segments, with domains $[0, t_0)$, $[t_0, t_1)$, and $[t_1, \ell)$, where ℓ is the end of timed trace. At each segment end point, such as t_0 and t_1, the trajectory may make a discontinuous jump, following the standard piecewise continuous trajectory model [15].

We specify this in Definition 5.1 by requiring that there is a strictly ordered sequence of real numbers I, that give the start and end point of each segment. $\mathbb{R}_{osq}$ is the subset of finite real sequences such that for every index n in the sequence less than its length minus one ($\#x - 1$), $x_n < x_{n+1}$. I must contain at least 0 and t, such that at least one segment is present, and only values between these two extremes. The timed trace f is required to be continuous on each interval $[I_n, I_{n+1})$, and convergent to a limit at I_{n+1}. The operator f *cont-on* A specifies that f is continuous on the range given by A, by requiring that, each point $t \in A$, the limit of $f(x)$ as x approaches t from above equals $f(t)$. We use the upper limit as the lower limit may be different, for example at the discontinuous jumps t_0 and t_1 in Fig. 1. The operator f *has-limit* k requires that there is a limit point l such that f converges toward l as it approaches k from below. Due, to discontinuity, the value at k may be different.

From this model, we can now introduce the core operators on timed traces, which we previously defined in [8], inspired by [27]. We reproduce them here for completeness and because our timed trace model has further constraints.

Definition 5.2 (Timed-trace Operators).

$$f \gg n \mathrel{\widehat{=}} \lambda\, x \bullet f(x - n) \qquad\qquad \varepsilon \mathrel{\widehat{=}} \emptyset$$

$$end(f) \mathrel{\widehat{=}} \min(\mathbb{R}_{\geq 0} \setminus \operatorname{dom}(f)) \qquad f \,\widehat{}\, g \mathrel{\widehat{=}} f \cup (g \gg end(f))$$

Function $f \gg n$ shifts the indices of a partial function $f : \mathbb{R}_{\geq 0} \nrightarrow A$ to the right by $n : \mathbb{R}_{\geq 0}$. The operator $\mathrm{end}(f)$ gives the end time of a trace $f : \mathbb{TT}_{\Sigma_c}$ by taking the infimum of the real numbers excluding the domain of f. The empty trace ε is the empty function. Finally, $f \frown g$ shifts the domain of g to start at the end of f, and takes the union. From these definitions, we prove the following theorem.

Theorem 5.3. *For any* Σ_c, $(\mathbb{TT}_{\Sigma_c}, \frown, \varepsilon)$ *forms a trace algebra.*

This model is the foundation for hybrid relations, which we now describe.

6 Hybrid Relations

In this section we describe our hybrid relational calculus, which specialises our theory of reactive relations with our timed trace model. We use this to give a denotational semantics to the hybrid programming language described in Sect. 2, including continuous variables, continuous specifications, and systems of ordinary differential equations (ODEs). A preliminary presentation of the materials in this section can be found in a previous technical report [28].

6.1 Continuous Variables

A hybrid relation is a specialised reactive relation where the underlying trace model is $(\mathbb{TT}_{\Sigma_c}, \frown, \varepsilon)$, with $tr, tr' : \mathbb{TT}_{\Sigma_c}$ and $st, st' : \Sigma_d \times \Sigma_c$. Intuitively, a hybrid relation describes a set of trajectories that characterise the possible behaviours of the continuous variables. The trace contribution (tt) refers to a particular evolution of the continuous state space, Σ_c. We introduce the syntax $\ell \triangleq \mathrm{end}(tt)$, which refers to the length of the present evolution in a hybrid relation [29].

As outlined in Sect. 4, our theory provides us with the operators of an imperative programming language. Consequently, we do not redefine them here, but reuse their existing definitions and laws. This is a key contribution of our approach – we need now only consider the specialised continuous evolution operators.

The hybrid state in st consists of the discrete and continuous state. We introduce independent lenses $d : \Sigma_d \Longrightarrow \Sigma$ and $c : \Sigma_c \Longrightarrow \Sigma$ for these subregions, respectively, which are the first and second projections. We introduce the syntax $s{:}x$ to project the part of state space s described by lens x, and then refer to discrete-state variables using $d{:}x$ and continuous-state variables using $c{:}y$.

Continuous variables are modelled as projections from the state space, Σ_c, over time. We likewise use lenses to model these projections, so that each variable x identifies a region of Σ_c, such as $x : \mathbb{R} \Longrightarrow \Sigma_c$. The source type of each lens, that is the type of data it refers to, is not limited to $\mathbb{R}$ but can be any topological space. A trajectory variable expression, $\tilde{x}(t)$, can then be defined as follows:

Definition 6.1 (Trajectory Variables). $\tilde{x}(t) \triangleq tt(t){:}x$

A trajectory variable, $\widetilde{x}$, is a function that obtains the continuous state space from the timed trace, recorded in tt, at time $t : \mathbb{R}_{\geq 0}$, and then projects the corresponding region using lens x. Here, t denotes time relative to the start of an evolution, and not absolute time, a property imposed by healthiness condition **R2**. Absolute time should instead be modelled as a distinguished continuous variable (cf. Sect. 2). It is also important to distinguish these trajectory variables, which are functions on the timed trace, from **state variables**, that is the valuation of the continuous variables at the start or end of a computation, characterised by st. These quantities are related, but are not identical. The value of $\widetilde{x}(t)$ is not *a priori* the same as the corresponding variable $c{:}x'$, for example, because the former is part of tt, but the latter is part of st, and $st \bowtie tt$. Later in this section, we will introduce coupling invariants to link these quantities.

Next, we describe instant relations, which lift relational predicates on the continuous state to hybrid relations:

Definition 6.2 (Instant Relations). $P @ t \triangleq [c' \mapsto tt(t)] \dagger P$ &

An instant predicate expression, $P @ t$, lifts primed continuous state variables, referred to in P, to continuous trajectory variables. P is a relation over the discrete and continuous state variables, that is a subset of $\mathbb{P}(\Sigma \times \Sigma)$.

To exemplify, the expression $(x' > 7.5) @ t$ is equivalent to $\widetilde{x}(t) > 7.5$, that is, the predicate that asserts that continuous state variable x is greater than 7.5 at time t. Instant relations can also refer to the initial values of continuous variables: primed variables (x') are used to denote the valuation of x at t, whereas its unprimed variant (x) simply refers to the initial value. Thus, the relation $(x > x') @ t$ is equivalent to $\widetilde{x}(t) > x$ — the value of x in the trajectory at time t is greater than it was initially. Effectively P is a relation between initial values of continuous variables, alternatively written as x_0, and the valuation of the variables at t. The definition of $P @ t$ simply substitutes the valuation of the continuous state variable c' for $tt(t)$: the trajectory state at t.

We next define an interval operator, inspired by Duration Calculus [29,30]:

Definition 6.3 (Interval). $dur[P(t)] \triangleq \textbf{R1}\,(\forall\, \tau \in [0, \ell) \bullet P(\tau) @ \tau)$ &

An interval specification $dur[P(t)]$ states that such a relation P holds over the entire evolution of the trajectory. Here, P is also parametrised by the current time t, which allows continuous variables to also depend on time. Technically, $t : \mathbb{R}_{\geq 0}$ is distinguished variable that is often used in continuous time predicates. We can obtain a Duration Calculus style specification operator with $\lceil p \rceil \triangleq dur[p']$, where p is a predicate on undashed variables only that does not mention t. This simplified operator states that the invariant p holds over the evolution.

The definition of $dur[P(t)]$ states that P holds at every instant τ between 0 and ℓ, and additionally enforces **R1** to ensure only healthy timed traces are permitted. The construction is automatically **R2** since it only refers to tt and not tr or tr' explicitly. Thus, since neither ok nor $wait$ are mentioned, $dur[P(\tau)]$ is an **RR** healthy reactive relation. Moreover, it is also **RC** healthy, since the set of timed traces specified is prefix-closed. The derived duration operator has the following laws, adapted from Duration Calculus, as theorems:

Theorem 6.4 (Interval Laws).

$$\lceil p \wedge q \rceil = (\lceil p \rceil \wedge \lceil q \rceil)$$
$$\lceil p \vee q \rceil \sqsubseteq (\lceil p \rceil \vee \lceil q \rceil)$$
$$\lceil p \rceil \,; \lceil p \rceil = \lceil p \rceil$$

$$\lceil \textbf{false} \rceil = (tr' = tr)$$
$$\lceil \textbf{true} \rceil = \textbf{true}_r$$

6.2 Continuous Function Evolution

The operators defined so far only permit specification of trajectory variables. In order to link these to continuous state variables so that, for instance, we can assign continuous variables, we define two coupling invariant operators.

Definition 6.5 (Continuous Coupling Invariants).

$$\textbf{II}_x \triangleq \textbf{R1}\,(c{:}x = \tilde{x}(0)) \qquad \textbf{rl}_x \triangleq \textbf{R1}\left(c{:}x' = \lim_{t \uparrow \ell} \tilde{x}(t) \right)$$

The first coupling invariant, $\textbf{II}_x$, links together the initial value of continuous state variable x with the corresponding trajectory variable at time 0. The second, $\textbf{rl}_x$, links the final value of continuous state variable x (that is, x') with the limit of the corresponding trajectory variable as it approaches the duration of the evolution (ℓ) from the left. By Definition 5.1, we know that the latter limit must exist, since our timed traces are piecewise convergent.

The asymmetry of the two invariants is important. Whilst the trajectory explicitly defines a value at time 0, as invoked by $\textbf{II}$, it does not define one at ℓ since the domain is the right-open interval $[0, \ell)$. The final value exists, however, because the timed trace converges to a limit. However, when sequentially composing hybrid relations, and thus composing the two trajectories, a discrete jump is permitted so that the value at t and the left limit at t need not be the same. Both $\textbf{II}_x$ and $\textbf{rl}_x$ are healthy reactive relations.

We can now define operators for continuous function evolution:

Definition 6.6 (Function Evolution).

$$\tilde{x}(t) \leftarrow f(t) \triangleq (dur\,[x' = f(t)] \wedge \ell > 0)$$

$$\tilde{x}(t) \underset{\leq d}{\leftarrow} f(t) \triangleq (\tilde{x}(t) \leftarrow f(t) \wedge \ell \leq d)$$

$$\tilde{x}(t) \underset{[s,d]}{\leftarrow} f(t) \triangleq (\tilde{x}(t) \leftarrow f(t) \wedge \ell \in [s, d] \wedge \textbf{d}' = \textbf{d} \wedge \textbf{rl}_{\textbf{v}})$$

where $x : \mathbb{R}^n \Longrightarrow \Sigma_c,\ f : \mathbb{R}_{\geq 0} \to \mathbb{R}^n,$ and $d : \mathbb{R}_{\geq 0}.$

Here, v is a special variable that denotes the entirety of the state space. The first operator, $\widetilde{x}(t) \leftarrow f(t)$, states that the trajectory variable x evolves according to continuous function f in n variables. We require that such an evolution have non-zero duration, as otherwise the function's behaviour cannot be observed. Lens x can consist of several continuous variables, and thus a function evolution can be used to encode a system of simultaneous algebraic equations, as present, for example, in Modelica. It is also worth noting that other continuous variables not mentioned in such a statement are unconstrained and thus behave nondeterministically. This is an important feature of the model as it allows the use of nondeterminism to model concurrency of parallel hybrid processes.

We exemplify the semantics of function evolution with the calculation below:

Example 6.7 (Evolution Calculation).

$$\widetilde{v}(t) \leftarrow v - 9.81 \cdot t$$
$$= (dur[v' = v - 9.81 \cdot t] \wedge \ell > 0)$$
$$= (\textbf{R1}(\forall\, t \in [0, \ell) \bullet \widetilde{v}(t) = v - 9.81 \cdot t) \wedge \ell > 0)$$
$$= (tr \leq tr' \wedge (\forall\, t \in [0, \ell) \bullet tt(t){:}v = v - 9.81 \cdot t) \wedge \ell > 0)$$

The evolution is first mapped to a duration with the equation $v' = v - 9.81 \cdot t$, meaning that v is assigned the initial value of v minus $9.81 \cdot t$, at each instant. In the second and third steps, the interval operator is expanded to a predicate that assigns a value to $\widetilde{v}$ over the whole duration. Ultimately, this continuous variable is simply a reference to tt, which shows how the semantics builds on generalised reactive relations. Function evolution also admits the following valuable theorem:

Theorem 6.8. $c{:}y := v \,;\, \widetilde{x}(t) \leftarrow f(t) \;=\; \widetilde{x}(t) \leftarrow (f(t))[v/y]$ ✿

This law shows the effect of pushing a leading assignment to y into a function evolution. Any instance of y in the continuous function expression $f(t)$ is replaced by v. This allows evaluation of any expressions that depend upon the initial state.

The second operator in Definition 6.6, $\widetilde{x}(t) \leftarrow_{\leq d} f(t)$, is the same as the above, but adds the requirement that the duration be at most d. The third and final operator, $\widetilde{x}(t) \leftarrow_{[s,d]} f(t)$, states that the evolution terminates nondeterministically in the interval $s \leq t \leq d$. This operator explicitly terminates the function's evolution and thus additionally states that all discrete variables should remain the same as they were at the start, and applies coupling invariant rl_v to set the final state of all continuous variables. All the function evolution operators in Definition 6.6 form healthy reactive relations.

6.3   Preemption of Evolution

We next define the preemption operator:

Definition 6.9. $P \,\triangle\, \langle b \,|\, c \rangle \;\triangleq\; (P \wedge dur[b] \wedge \ell > 0 \wedge rl_v \wedge c' \wedge d' = d)$ ✿

The preemption operator $(P \bigtriangleup \langle b \mid c \rangle)$ states that P evolves for some non-zero duration, while condition b holds. At some undetermined point, c should become true finally, and at this point the operator can terminate. This yields final values for all variables, obtained using the right limit, and requires that all discrete variables remain unchanged over the evolution.

Intuitively, the first condition, b, is similar to the invariants present in hybrid automata [31]. Evolution of P can continue while b remains true, which is ensured by conjunction with the interval specification $dur[b]$. On the other hand, evolution of P can terminate whenever the final continuous state satisfies c. Since b and c can overlap there is potential nondeterminism as to when P terminates, which is necessary when handling numerical imprecision. In the special case that $b = (\neg c)$, there is at most one instant at which P terminates, leading to a precise and purely deterministic preemption. If c never becomes true then this operator evaluates to **false**, that is, a non-terminating reactive relation.

We give an important theorem regarding termination of a function evolution:

Theorem 6.10. *We assume that f is a continuous function on the domain $[0, l]$, where $l > k$ and $k > 0$, and the following conditions hold:*

1. *b is satisfied for all instants $t \in [0, l)$: $\forall t \in [0, l) \bullet b[f(t)/x']$;*
2. *b becomes false at l: $\neg b[f(l)/x']$;*
3. *c is not satisfied for all instants $t \in [0, k)$: $\forall t \in [0, k) \bullet \neg c[f(t)/x']$;*
4. *c becomes true at k and stays true until l: $\forall t \in [k, l) \bullet c[f(t)/x']$.*

Then the following equality holds:

$$(\widetilde{x}(t) \leftarrow f(t) \bigtriangleup \langle b \mid c \rangle) = \left(\widetilde{x}(t) \underset{[k..l]}{\leftarrow} f(t) \right)$$

This theorem shows the conditions under which a function evolution, with a given invariant and preemption condition, will terminate. The first two assumptions ensure that the invariant b is true initially, and remains true until l. The remaining two assumptions state that c was not true for some period, until k at which point it becomes true and stays true until l. This being the case, the preemption will occur nondeterministically at some point between k and l. A special case is when $k = l$, in which case there is precisely one instant when this occurs. This theorem is useful in languages like Modelica where the evolution of a differential equation can be halted when a specific condition is reached.

6.4   Derivatives and Ordinary Differential Equations

The ability to express derivatives of continuous variables is central to hybrid system modelling. In the hybrid relational calculus we introduce the notation x **has-der** $f(t)$ which states that the derivative of continuous variable x is determined by expression f, which is parametrised over time t. This is equivalent to the usual calculus notation $\dot{x}(t) = f(t, x)$. For example, we can write constraints like x **has-der** $2 \cdot x$, which states that x is changing at the rate of $2 \cdot x$.

A system of ODEs, $\dot{x}(t) = f(t, x(t))$, specifies a family of continuous solution functions, $x : \mathbb{R}_{\geq 0} \to \mathbb{R}^n$, that specify the value for the n variables at each instant. The system is defined by function $f : \mathbb{R}_{\geq 0} \times \mathbb{R}^n \to \mathbb{R}^n$ that gives the derivative of each variable at time t, and depends on the initial value, that is $x(t)$. A solution is any function x that changes at the rate specified by f.

Naturally, when animating or verifying a system, a single solution is normally desired. For this, it is necessary to construct an initial value problem (IVP) that supplements the system of ODEs with initial values for all continuous variables. Then the Picard-Lindelöf theorem [32] can be applied to show that, provided f is Lipschitz continuous, a unique solution exists to the initial value problem [33]. Lipschitz continuity essentially limits the rate at which a continuous function can change. We now describe our operator for systems of ODEs:

Definition 6.11. $\langle \dot{x}(t) \bullet f(t, x) \rangle \triangleq (II_x \wedge x \; \textbf{\textit{has-der}}(f(t)(x)))$

The operator takes two parameters: $x : \mathbb{R}^n \Longrightarrow \Sigma_c$, which is a lens projecting a vector of reals from the continuous state; and f, the ODE specification function described above. The definition applies the initial value coupling invariant, and asserts that lens x has the derivative given by the characteristic ODE function f. It does not apply the final state coupling invariant, $\textbf{\textit{rl}}$, as a system of ODEs only produces a final value when it is preempted. Usually, though not necessarily, ODEs are guarded by the $\triangle \langle b \,|\, c \rangle$ operator. Every operator of which the ODE operator is composed is $\textbf{\textit{R1}}$ and $\textbf{\textit{R2}}$, and thus it is a healthy reactive relation.

In order to solve differential equations, it is necessary to set up an IVP. The following theorem shows how a solution may be used to transform an ODE to symbolic solution function evolution.

Theorem 6.12. *If, for any $v : \mathbb{R}^n$ and $l > 0$, $g(v)$ is the unique solution to f on the interval $[0, l]$, and $g(v)(0) = v$ then*

$$\langle \dot{x}(t) \bullet f(t, x) \rangle = \widetilde{x}(t) \leftarrow g(x)(t)$$

This theorem allows us to transform a differential equation into a solution function evolution. It has some subtleties that require further explanation. Function $g : \mathbb{R}^n \to \mathbb{R} \to \mathbb{R}^n$ is the solution function, but it depends on the initial value for variables which is why it has two inputs. This allows us to abstract from IVPs when symbolically solving an ODE. Thus, we require that for any given initial valuation of the continuous state v, $g(v)$ is the unique solution to f. Moreover, we require that the function's value at time 0 be the initial value we have supplied; a kind of sanity check for the function. If all these conditions are satisfied then the ODE can be rewritten to $\widetilde{x}(t) \leftarrow g(x, t)$. The x on the right hand side of the arrow is the initial value of x, as usual for the relational calculus. Thus, the solution function is fully described when an initial value is supplied by a preceding assignment, for example by use of Theorem 6.8.

In terms of showing that a function is a unique solution, it suffices to show that the function is a solution and then to exhibit an appropriate Lipschitz constant. In Isabelle/HOL the former of these two can be accomplished through a tactic we have written called **ode-cert** that certifies a solution to an ODE by applying derivative introduction rules.

To exemplify, we give the following calculation of the first step of Example 2.2:

Example 6.13. (ODE Calculation).

$$h, v := 2, 0 \; ; \; \langle (\dot{h}(t), \dot{v}(t)) \bullet (v, -g) \rangle$$

$$= h, v := 2, 0 \; ; \; \begin{pmatrix} h(t) \\ v(t) \end{pmatrix} \leftarrow \begin{pmatrix} v \cdot t - g \cdot t^2/2 + h \\ v - g \cdot t \end{pmatrix} \tag{6.12}$$

$$= \begin{pmatrix} h(t) \\ v(t) \end{pmatrix} \leftarrow \begin{pmatrix} 0 \cdot t - g \cdot t^2/2 + 2 \\ 0 - g \cdot t \end{pmatrix} \tag{6.8}$$

We first obtain the unique solution to the ODEs, which can be done using a typical computer algebra tool like Mathematica, and then rewrite this to a function evolution. We also push forward the assignment using Theorem 6.8 to set initial values for continuous variables.

6.5 Hybrid Reactive Contracts

Whilst hybrid relations can be used to model programs, they do not allow us to distinguish terminating, non-terminating, and divergent behaviours[5]. Specifically, a dynamically evolving ODE, $\langle \dot{x}(t) \bullet f(t) \rangle$ can continue indefinitely. In spite of this, it does not satisfy the following theorem [2]:

$$\langle \dot{x}(t, x) \bullet f(t) \rangle \; ; \; P \; = \; \langle \dot{x}(t, x) \bullet f(t) \rangle$$

For example, if P is an assignment, $x := v$, then the results of it are observable in such a composition. This is the reason that we often place ODEs in the context of a preemption operator[6], which correctly handles termination. This issue is analogous to the well-known problem with basic relational model of programs, which motivated the UTP theory of designs [7,34].

Our solution, similarly, is to introduce a UTP theory of reactive contractual specifications, and use the *ok* and *wait* observational variables to distinguish divergent and intermediate observations. This approach captures non-termination in reactive systems in a way that avoids complex reasoning associated with infinite traces. In previous work [25,26,35] we developed a UTP theory of generalised reactive designs, building on our theory of reactive relations [8] (Sect. 4) and prior work with *Circus* [16,17]. We use this to develop a

---

[5] A concept capturing erroneous behaviours such as unproductive non-termination.
[6] This is also true of d$\mathcal{L}$ hybrid programs, which are modelled similarly.

contract notation, and a method for automatically calculating the semantics of reactive programs for the purpose of verification [26]. As for designs, our reactive contracts also support refinement with assume/guarantee style reasoning [36,37]. It allows a unified set of laws for a diverse set of languages, including CSP [7], *Circus* [17], timed extensions [38], and of course our hybrid relations.

A reactive contract, $[\,P_1 \vdash P_2 \mid P_3\,]$, consists of three reactive relations that specify the (1) assumption, P_1, and (2) guarantee for intermediate observations, P_2, and terminating observations, P_3. Assumption P_1 is a reactive condition (Definition 4.2): it can refer only the initial state (st) and trace (tt), and the characterised set of traces must be prefix closed. If the assumption of a contract is violated, then the result is the most nondeterministic reactive designs, called **Chaos** $\triangleq [\,$**false** $\vdash$ **false** $\mid$ **false**$\,]$, which corresponds to divergent behaviour. P_2 characterises the intermediate or waiting observations of the reactive program; consequently it is a reactive relation that, like P_1, refers only to st and tt. P_3 characterises terminating observations, and so can refer to st, tt, and also st'.

We can now lift our ODE operator so that it is correctly non-terminating:

Definition 6.14. $\langle\!\langle\, \dot{x}(t) \bullet f(t,x) \,\rangle\!\rangle \triangleq [\,$**true**$_r \vdash \langle\, \dot{x}(t) \bullet f(t,x) \,\rangle \mid$ **false**$\,]$

The assumption of the lifted ODE is true, since there is no divergent behaviour. The terminating guarantee is **false**, since this is a non-terminating operator. The intermediate guarantee is simply our hybrid relational ODE operator, so that evolutions of the ODE are flagged as intermediate observations. Then, we can use the following contract theorems for reasoning about compositions [25,26]:

Theorem 6.15 (Reactive Design Laws).

$$[\,P_1 \vdash P_2 \mid P_3\,]\,;[\,\textbf{true}_r \vdash Q_2 \mid Q_3\,] \;=\; [\,P_1 \vdash P_2 \vee P_3 \,;\, Q_2 \mid P_3 \,;\, Q_3\,] \quad (6.15.1)$$

$$\textbf{Chaos} \sqcap [\,P_1 \vdash P_2 \mid P_3\,] \;=\; \textbf{Chaos} \quad\quad\quad (6.15.2)$$

$$[\,P_1 \vdash P_2 \mid \textbf{false}\,]\,;[\,Q_1 \vdash Q_2 \mid Q_3\,] \;=\; [\,P_1 \vdash P_2 \mid \textbf{false}\,] \quad\quad (6.15.3)$$

$$\textbf{Chaos}\,;[\,P_1 \vdash P_2 \mid P_3\,] \;=\; \textbf{Chaos} \quad\quad\quad (6.15.4)$$

$$[\,\textbf{false} \vdash P_2 \mid P_3\,] \;=\; \textbf{Chaos} \quad\quad\quad (6.15.5)$$

(6.15.1) is the basic law for sequential composition[7]. When composing contracts in the sequence, an intermediate observation is either an intermediate observation of the first contract (P_2), or a terminating observation of the first, followed by an intermediate observation of the second ($P_3 \,;\, Q_2$). A terminating observation requires that both contracts can terminate $P_3 \,;\, Q_3$. (6.15.2) show that

---

[7] For brevity, we present a simplified law where the second assumption is **true**$_r$.

Chaos is indeed the most nondeterministic reactive contract. The remaining laws are essentially corollaries of (6.15.1). Of particular interest for ODEs is (6.15.3), which has the following law as a consequence:

Theorem 6.16. $\langle\!\langle \dot{x}(t) \bullet f(t) \rangle\!\rangle \;;\; P = \langle\!\langle \dot{x}(t) \bullet f(t) \rangle\!\rangle$

Similar results can be achieved for the function evolution operators.

A further advantage of hybrid reactive contracts is to encode assumptions about continuous variables outside of the system's control (e.g. monitored variables). Assumptions can, for example, be specified using the interval operator of Definition 6.3, since this constructs reactive conditions. For example, we can specify a division block for the control law languages of Modelica or Simulink:

Example 6.17. $\mathrm{Div}(x, y, z) \triangleq [\, \lceil y \neq 0 \rceil \vdash \lceil z = x/y \rceil \mid \textit{false}\,]$

We encode a division block with two inputs, x and y, and a single output z. These are all modelled as lenses into the continuous state ($\mathbb{R} \Longrightarrow \Sigma_c$), that correspond to connections in a block diagram, and are given as parameters. The intuition here is that every wire in a control law diagram is modelled as a lens. The divison block is a non-terminating hybrid process that in every intermediate state requires that the continuous variable z take the value of x/y. The assumption requires that $y \neq 0$, to ensure that division by zero cannot occur. Using a pattern like this, we can give semantics to a large number of blocks in the Simulink and Modelica block libraries[8]. This allows us to use hybrid reactive designs to reason about control law diagrams.

7   Mechanisation and Example

The hybrid relational calculus, and the theorems described in Sect. 6 are mechanised in Isabelle/UTP. For this, we employ Isabelle's implementation of multivariate analysis [39], including its symbolic real numbers, Euclidean spaces, limits, and derivatives. We also utilise Immler's library for ODEs and IVPs [33,40], which allows us to certify that a function is the solution to a system of ODEs.

In order to exemplify the use of the mechanisation, we describe part of a tram model, which is part of a previous industrial case study [14]. We reproduce it here for the purposes of illustration, with adaptation for our new hybrid relational model. We focus on the situation when the tram is slowing due to an approaching red signal, and formalise this using variables for acceleration *acc*, velocity *vel*, and track position *pos*. We note that *normal-deceleration* below is negative and determines the rate at which the tram reduces its speed as the brakes are applied.

---

[8] See https://build.openmodelica.org/Documentation/Modelica.Blocks.html.

Definition 7.1 (Braking Tram in Hybrid Relational Calculus).

$$BrakingTrain \triangleq \left(\begin{array}{l} acc, vel, pos := normal\text{-}deceleration, max\text{-}speed, 0 \; ; \\ \left\langle \begin{pmatrix} \dot{acc} \\ \dot{vel} \\ \dot{pos} \end{pmatrix} \bullet \begin{pmatrix} 0 \\ \dot{acc} \\ \dot{vel} \end{pmatrix} \right\rangle \triangle \; \langle vel > 0 \,|\, vel \leq 0 \rangle \; ; \\ acc := 0 \end{array} \right)$$

We assign initial values to the continuous variables, and then evolve them until the velocity reaches 0. In this instance, we do not allow non-determinism here, but record the precise instant that the velocity is 0. Thus, the evolution invariant is $vel > 0$, and the preemption condition is $vel \leq 0$. After this, we set the acceleration to 0, so that the tram halts and does not start moving backwards. Though this model is highly idealised, a more realistic model, which, for example, introduces pertubations into the acceleration due to external influences like weather, can be described by adding periodic preemption conditions and non-deterministic assignments to corresponding variables.

This example is encoded in Isabelle/UTP, as shown in Fig. 2, where the preemption operator has the syntax $P \; \boldsymbol{inv} \; b \; \boldsymbol{until}_h \; c$. We also mechanise a proof that the train stops before the end of the track, that is,

Theorem 7.2. $(accl' = 0 \wedge dur[pos < 44]) \sqsubseteq BrakingTrain$ ♠

holds, where $44m$ is the track length. The specification to the left states that, for all possible evolutions, the final value of the acceleration is 0 and pos is always less than 44. This should then be refined by our hybrid relation, $BrakingTrain$. For the sake of brevity, we elide details of the proof in Isabelle, other than the first four steps. The proof proceeds as follows:

1. Solve the ODE to obtain a function evolution statement (Theorem 6.12);
2. Use the assigned values to obtain the initial conditions (Theorem 6.8);
3. Calculate the time at which the velocity reaches zero (Theorem 6.10);
4. Finally, prove that the position at every earlier instant is less than 44 m.

The final step requires that we solve a polynomial inequality:

$$(104/25) \cdot t - (7/10) \cdot t^2 < 44$$

which includes the position derivative solution. In Isabelle, this can be done using the approximate tactic [41], which applies floating-point computation.

```
definition "BrakingTrain =
    (c:accel, c:vel, c:pos) :=ᵣ («normal_deceleration», «max_speed», «0») ;;
    ({&accel,&vel,&pos} • train_ode(ti))ₕ inv ¬$vel´≤ᵤ0 untilₕ ($vel´≤ᵤ0) ;; c:accel :=ᵣ 0"

theorem braking_train_pos_le:
    "($st:c:accel´ =ᵤ 0 ∧ ⌈$pos´ <ᵤ 44⌉ₕ) ⊑ BrakingTrain" (is "?lhs ⊑ ?rhs")
proof -
    — ‹ Solve ODE, replacing it with an explicit solution: @{term train_sol}. ›
    have "?rhs =
        (c:accel, c:vel, c:pos) :=ᵣ («-1.4», «4.16», «0») ;;
        {&accel,&vel,&pos} ←ₕ «train_sol($accel,$vel,$pos)ₐ(«ti»)ₐ untilₕ ($vel´ ≤ᵤ 0) ;;
        c:accel :=ᵣ 0"
    by (simp only: BrakingTrain_def train_sol)
    — ‹ Set up initial values for the ODE solution using assigned variables. ›
    also have "... =
        {&accel,&vel,&pos}←ₕ«train_sol(-1.4,4.16,0)(ti)» untilₕ ($vel´≤ᵤ0) ;; c:accel :=ᵣ 0"
    by (rel_auto)
    — ‹ Find the point at which the train stops ›
    also have "... =
        (({&accel,&vel,&pos} ←ₕ(«416/140») «train_sol(-1.4,4.16,0)(ti)»)) ;; c:accel :=ᵣ 0"
```

Fig. 2. The braking tram in Isabelle/UTP

8 Conclusions and Discussion

We have described our UTP theory of hybrid relations that specialises reactive relations with a continuous timed trace model. A key result is the unification of hybrid models [1,2,13] and reactive programs [26], through our generalised theories of reactive relations and reactive designs. In a parallel development, we have used the generalised theory to mechanise a semantics and verification tool for *Circus* [17,26] (and thus CSP [42]), and our hybrid theory shares many of the laws, such as those in Theorem 6.15. This, we believe, shows the immense and practical value of unification. Our theory can also be used as a foundation for automated verification tools for hybrid programs in Isabelle/UTP [10], and we plan to apply it to verification of Modelica dynamical models, by extending our previous semantics [13] that used an early version of our UTP hybrid theory.

The two most related works are differential dynamic logic [1,4] (d$\mathcal{L}$), and HCSP [2,3,5], both of which have substantially influenced our direction.

Our model is more expressive than standard d$\mathcal{L}$ hybrid programs, since we encode an explicit trajectory, whilst d$\mathcal{L}$ encodes the initial/intermediate value pairs for each variable in a binary relation. This allows us to separate ODEs from preemption, which in d$\mathcal{L}$ are combined in a single operator, $\{x' = \theta \,\&\, b\}$, where b is the boundary condition. This can be useful when constructing systems by composition of continuous and discrete components, where b is not known *a priori*. An explicit trace model is also a prerequisite for modelling networks of communicating hybrid systems [2]. There is also a d$\mathcal{L}$ extension called dTL^2 [43] that also employs an explicit trajectory and is similar to our model.

Proof support in d$\mathcal{L}$'s tool, KeYmaera X [4], is clearly far more advanced than our implementation in Isabelle/UTP. Nevertheless, we are currently working on implementing d$\mathcal{L}$ in Isabelle/UTP[9], based on a recent implementation of d$\mathcal{L}$ in Isabelle/HOL [44], and hope to report on this soon. This will allow to formally link the two theories, and also extensions like [43], via Galois connections, and harness the differential induction reasoning technique.

HCSP [2,3] models communicating hybrid systems using CSP-style process algebraic operators. There are two main denotational semantic models, the original one by He [2], which employs a UTP-style relational calculus, and a later one by Zhou [3], that employs Duration Calculus [29,30]. Our model is comparable to, though less expressive than [2]—since [2] models a more sophisticated form of trajectory based on super-dense time [15,45]—and is likely of equivalent expressivity with [3]. The semantics and algebraic laws in [2] are a strong inspiration for our work, and we believe that [2] is very similar to our reactive designs. For super-dense time [15,45], the trajectory has type $\mathbb{R}_{\geq 0} \times \mathbb{N} \nrightarrow \Sigma_c$ – the time domain is extended with a natural number that allows state changes that are "simultaneous-but-ordered". This is, arguably, needed to allow CSP-style events that are often interpreted to take a zero time duration. We hope in the future to explore whether such a trajectory model forms a trace algebra [8], so that our reactive designs hierarchy can be reused. Moreover, we will also explore weakening the trace algebra to support infinite traces which are at present forbidden.

In conclusion, the UTP approach has been an invaluable tool in this development. Whilst several hybrid computational theories exist, there are links between them, which UTP theories allow us to explore. Moreover, UTP allows us to link to theories that at first sight seem unrelated, such as *Circus* [17], as our reactive design theory shows. Our overarching message is this: *the UTP works*—it can capture languages of differing and heterogeneous paradigms and use the associated theories to develop and integrate verification tools. As Hoare and He reflected in the first chapter of the UTP book, when considering all the tools and artefacts that software engineering research is producing:

> "...to ensure that [analysis] tools may be safely used in combination, it is essential that these [underlying] theories be unified..." [7, page 21]

We believe that our hierarchy of theories and verification tools in Isabelle/UTP is evidence that UTP supports a practical approach for integration of formal analysis tools [46]. As systems become more complex in nature, as is the case with cyber-physical systems and autonomous robots, there is an even greater need to consider integration of heterogeneous computational paradigms [6]. The UTP allows us to approach one of the grand challenges for software engineering: integration of formal methods [6,46,47] for assurance of large-scale systems.

Acknowledgments. This work is funded by the CyPhyAssure project (CyPhyAssure: https://www.cs.york.ac.uk/circus/CyPhyAssure/), EPSRC grant EP/S001190/1. We would like to thank the anonymous reviewers for their thorough and helpful input, which has improved the presentation of our work.

---

[9] Differential dynamic logic in Isabelle/UTP 🜚.

References

1. Platzer, A.: Differential dynamic logic for hybrid systems. J. Autom. Reasoning **41**, 143–189 (2008)
2. He, J.: From CSP to hybrid systems. In: Roscoe, A.W. (ed.) A Classical Mind: Essays in Honour of C. A. R. Hoare, pp. 171–189. Prentice Hall, Upper Saddle River (1994)
3. Chaochen, Z., Ji, W., Ravn, A.P.: A formal description of hybrid systems. In: Alur, R., Henzinger, T.A., Sontag, E.D. (eds.) HS 1995. LNCS, vol. 1066, pp. 511–530. Springer, Heidelberg (1996). https://doi.org/10.1007/BFb0020972
4. Fulton, N., Mitsch, S., Quesel, J.-D., Völp, M., Platzer, A.: KeYmaera X: an axiomatic tactical theorem prover for hybrid systems. In: Felty, A.P., Middeldorp, A. (eds.) CADE 2015. LNCS (LNAI), vol. 9195, pp. 527–538. Springer, Cham (2015). https://doi.org/10.1007/978-3-319-21401-6_36
5. Wang, S., Zhan, N., Zou, L.: An improved HHL prover: an interactive theorem prover for hybrid systems. In: Butler, M., Conchon, S., Zaïdi, F. (eds.) ICFEM 2015. LNCS, vol. 9407, pp. 382–399. Springer, Cham (2015). https://doi.org/10.1007/978-3-319-25423-4_25
6. Gleirscher, M., Foster, S., Woodcock, J.: New opportunities for integrated formal methods. ACM Comput. Surv. (2019). https://arxiv.org/abs/1812.10103. Accepted subject to minor revision
7. Hoare, C.A.R., He, J.: Unifying Theories of Programming. Prentice-Hall, Upper Saddle River (1998)
8. Foster, S., Cavalcanti, A., Woodcock, J., Zeyda, F.: Unifying theories of time with generalised reactive processes. Inf. Process. Lett. **135**, 47–52 (2018)
9. Foster, S., Zeyda, F., Nemouchi, Y., Ribeiro, P., Wolff, B.: Isabelle/UTP: mechanised theory engineering for unifying theories of programming. Arch. Formal Proofs (2019). https://www.isa-afp.org/entries/UTP.html
10. Foster, S., Baxter, J., Cavalcanti, A., Woodcock, J., Zeyda, F.: Unifying semantic foundations for automated verification tools in Isabelle/UTP, March 2019. https://arxiv.org/abs/1905.05500. Submitted to Science of Computer Programming
11. Foster, S., Zeyda, F., Woodcock, J.: Unifying heterogeneous state-spaces with lenses. In: Sampaio, A., Wang, F. (eds.) ICTAC 2016. LNCS, vol. 9965, pp. 295–314. Springer, Cham (2016). https://doi.org/10.1007/978-3-319-46750-4_17
12. Foster, S., Zeyda, F., Woodcock, J.: Isabelle/UTP: a mechanised theory engineering framework. In: Naumann, D. (ed.) UTP 2014. LNCS, vol. 8963, pp. 21–41. Springer, Cham (2015). https://doi.org/10.1007/978-3-319-14806-9_2
13. Foster, S., Thiele, B., Cavalcanti, A., Woodcock, J.: Towards a UTP semantics for modelica. In: Bowen, J.P., Zhu, H. (eds.) UTP 2016. LNCS, vol. 10134, pp. 44–64. Springer, Cham (2017). https://doi.org/10.1007/978-3-319-52228-9_3
14. Zeyda, F., Ouy, J., Foster, S., Cavalcanti, A.: Formalising cosimulation models. In: Cerone, A., Roveri, M. (eds.) SEFM 2017. LNCS, vol. 10729, pp. 453–468. Springer, Cham (2018). https://doi.org/10.1007/978-3-319-74781-1_31
15. Lee, E.A.: Constructive models of discrete and continuous physical phenomena. IEEE Access **2**, 797–821 (2014)
16. Cavalcanti, A., Woodcock, J.: A tutorial introduction to CSP in *unifying theories of programming*. In: Cavalcanti, A., Sampaio, A., Woodcock, J. (eds.) PSSE 2004. LNCS, vol. 3167, pp. 220–268. Springer, Heidelberg (2006). https://doi.org/10.1007/11889229_6

17. Oliveira, M., Cavalcanti, A., Woodcock, J.: A UTP semantics for circus. Formal Aspects Comput. **21**, 3–32 (2009)
18. Back, R.J., Wright, J.: Refinement Calculus: A Systematic Introduction. Springer, Berlin (1998)
19. Dijkstra, E.W.: Guarded commands, nondeterminacy and formal derivation of programs. Commun. ACM **18**(8), 453–457 (1975)
20. Feliachi, A., Gaudel, M.-C., Wolff, B.: Unifying theories in isabelle/HOL. In: Qin, S. (ed.) UTP 2010. LNCS, vol. 6445, pp. 188–206. Springer, Heidelberg (2010). https://doi.org/10.1007/978-3-642-16690-7_9
21. Feliachi, A., Gaudel, M.-C., Wolff, B.: Isabelle/*Circus*: a process specification and verification environment. In: Joshi, R., Müller, P., Podelski, A. (eds.) VSTTE 2012. LNCS, vol. 7152, pp. 243–260. Springer, Heidelberg (2012). https://doi.org/10.1007/978-3-642-27705-4_20
22. Boulton, R., Gordon, A., Gordon, M., Harrison, J., Herbert, J., van Tassel, J.: Experience with embedding hardware description languages in HOL. In: Proceedings of IFIP International Conference on Theorem Provers in Circuit Design, pp. 129–156 (1993)
23. Gomes, V.B.F., Struth, G.: Modal Kleene algebra applied to program correctness. In: Fitzgerald, J., Heitmeyer, C., Gnesi, S., Philippou, A. (eds.) FM 2016. LNCS, vol. 9995, pp. 310–325. Springer, Cham (2016). https://doi.org/10.1007/978-3-319-48989-6_19
24. Foster, J.: Bidirectional programming languages. Ph.D., thesis, University of Pennsylvania (2009)
25. Foster, S., Cavalcanti, A., Canham, S., Woodcock, J., Zeyda, F.: Unifying theories of reactive design contracts, December 2017. https://arxiv.org/abs/1712.10233. Under revision for Theoretical Computer Science
26. Foster, S., Ye, K., Cavalcanti, A., Woodcock, J.: Calculational verification of reactive programs with reactive relations and kleene algebra. In: Desharnais, J., Guttmann, W., Joosten, S. (eds.) RAMiCS 2018. LNCS, vol. 11194, pp. 205–224. Springer, Cham (2018). https://doi.org/10.1007/978-3-030-02149-8_13
27. Höfner, P., Möller, B.: An algebra of hybrid systems. J. Logic Algebraic Program. **78**(2), 74–97 (2009)
28. Cavalcanti, A., Foster, S., Thiele, B., Woodcock, J., Zeyda, F.: Final Semantics of Modelica. Technical report, INTO-CPS Deliverable, D2.3b, December 2017
29. Chaochen, Z., Ravn, A.P., Hansen, M.R.: An extended duration calculus for hybrid real-time systems. In: Grossman, R.L., Nerode, A., Ravn, A.P., Rischel, H. (eds.) HS 1991-1992. LNCS, vol. 736, pp. 36–59. Springer, Heidelberg (1993). https://doi.org/10.1007/3-540-57318-6_23
30. Zhou, C., Hoare, C.A.R., Ravn, A.P.: A calculus of durations. Inf. Process. Lett. **40**(5), 269–276 (1991)
31. Henzinger, T.A. In: The theory of hybrid automata, pp. 278–292. IEEE (1996)
32. Coddington, E.A., Levinson, N.: Theory of Ordinary Differential Equations. McGraw-Hill, New York (1955)
33. Immler, F., Hölzl, J.: Numerical analysis of ordinary differential equations in Isabelle/HOL. In: Beringer, L., Felty, A. (eds.) ITP 2012. LNCS, vol. 7406, pp. 377–392. Springer, Heidelberg (2012). https://doi.org/10.1007/978-3-642-32347-8_26
34. Woodcock, J., Cavalcanti, A.: A tutorial introduction to designs in unifying theories of programming. In: Boiten, E.A., Derrick, J., Smith, G. (eds.) IFM 2004. LNCS, vol. 2999, pp. 40–66. Springer, Heidelberg (2004). https://doi.org/10.1007/978-3-540-24756-2_4

35. Foster, S., Baxter, J., Cavalcanti, A., Miyazawa, A., Woodcock, J.: Automating verification of state machines with reactive designs and Isabelle/UTP. In: Bae, K., Ölveczky, P.C. (eds.) FACS 2018. LNCS, vol. 11222, pp. 137–155. Springer, Cham (2018). https://doi.org/10.1007/978-3-030-02146-7_7

36. Meyer, B.: Applying "design by contract". IEEE Comput. **25**(10), 40–51 (1992)

37. Benveniste, A., Caillaud, B., Ferrari, A., Mangeruca, L., Passerone, R., Sofronis, C.: Multiple viewpoint contract-based specification and design. In: de Boer, F.S., Bonsangue, M.M., Graf, S., de Roever, W.-P. (eds.) FMCO 2007. LNCS, vol. 5382, pp. 200–225. Springer, Heidelberg (2008). https://doi.org/10.1007/978-3-540-92188-2_9

38. Sherif, A., Cavalcanti, A., He, J., Sampaio, A.: A process algebraic framework for specification and validation of real-time systems. Formal Aspects Comput. **22**(2), 153–191 (2010)

39. Harrison, J.: A HOL theory of euclidean space. In: Hurd, J., Melham, T. (eds.) TPHOLs 2005. LNCS, vol. 3603, pp. 114–129. Springer, Heidelberg (2005). https://doi.org/10.1007/11541868_8

40. Immler, F.: Formally verified computation of enclosures of solutions of ordinary differential equations. In: Badger, J.M., Rozier, K.Y. (eds.) NFM 2014. LNCS, vol. 8430, pp. 113–127. Springer, Cham (2014). https://doi.org/10.1007/978-3-319-06200-6_9

41. Hölzl, J.: Proving inequalities over reals with computation in Isabelle/HOL. In: Proceedings of 2009 International Workshop on Programming Languages for Mechanized Mathematics Systems (PLMMS), pp. 38–45. ACM, August 2009

42. Brookes, S.D., Hoare, C.A.R., Roscoe, A.W.: A theory of communicating sequential processes. J. ACM **31**(3), 560–599 (1984)

43. Jeannin, J.-B., Platzer, A.: dTL^2: differential temporal dynamic logic with nested temporalities for hybrid systems. In: Demri, S., Kapur, D., Weidenbach, C. (eds.) IJCAR 2014. LNCS (LNAI), vol. 8562, pp. 292–306. Springer, Cham (2014). https://doi.org/10.1007/978-3-319-08587-6_22

44. Huerta y Munive, J.J., Struth, G.: Verifying hybrid systems with modal kleene algebra. In: Desharnais, J., Guttmann, W., Joosten, S. (eds.) RAMiCS 2018. LNCS, vol. 11194, pp. 225–243. Springer, Cham (2018). https://doi.org/10.1007/978-3-030-02149-8_14

45. Manna, Z., Pnueli, A.: Verifying hybrid systems. In: Grossman, R.L., Nerode, A., Ravn, A.P., Rischel, H. (eds.) HS 1991-1992. LNCS, vol. 736, pp. 4–35. Springer, Heidelberg (1993). https://doi.org/10.1007/3-540-57318-6_22

46. Paige, R.F.: A meta-method for formal method integration. In: Fitzgerald, J., Jones, C.B., Lucas, P. (eds.) FME 1997. LNCS, vol. 1313, pp. 473–494. Springer, Heidelberg (1997). https://doi.org/10.1007/3-540-63533-5_25

47. Galloway, A.J., Stoddart, B.: Integrated formal methods. In: Proceedings of INFORSID, INFORSID (1997)

Concurrency

The Inner and Outer Algebras
of Unified Concurrency

Andrew Butterfield(✉) [iD]

Trinity College Dublin, Dublin, Ireland
butrfeld@tcd.ie
http://www.scss.tcd.ie/

Abstract. Algebras have always played a critical role in Unifying Theories of Programming, especially in their role in providing the "laws" of programming. The algebraic laws form a triad with two other forms, namely operational and denotational semantics. In this paper we demonstrate that algebras are not just for providing external laws for reasoning about programs. In addition, they can be very beneficial for assisting in the development of theoretical models, most notably denotational semantics. We refer to the algebras used to develop a denotational model as "inner algebras", while the resulting algebraic semantics we consider to be an "outer algebra". In this paper we present a number of inner algebras that arose in the development of a fully compositional denotational semantics, called UTCP, for shared-state concurrency. We explore how these algebras helped to develop (and debug!) the theory, and discuss how they may assist in the ultimate aim of exposing the outer algebra of UTCP, which we expect to be very similar to Concurrent Kleene Algebra.

Keywords: Unifying Theories of Programming · Inner algebras ·
Outer algebras · Shared-variable concurrency ·
Concurrent Kleene Algebras

1 Introduction

The work reported here has been inspired by the "Views" paper [9], which describes how a range of approaches to reasoning about shared-variable concurrency can be mapped down onto instantiations of commutative semi-groups and monoids. The paper introduced a simple language of syntactic commands, and used it as a baseline to connect a wide variety of formal approaches to concurrency. Approaches covered in [9] include various Separation logics [8], type-theories, Owicki-Gries [20], and Rely-Guarantee [17], among others. Our intention in developing a UTP semantics of this command language is to be able to use it as a foundation on which to build UTP theories of the above approaches

This work was supported, in part, by Science Foundation Ireland grant 13/RC/2094 to Lero - the Irish Software Engineering Research Centre (www.lero.ie).

© Springer Nature Switzerland AG 2019
P. Ribeiro and A. Sampaio (Eds.): UTP 2019, LNCS 11885, pp. 157–175, 2019.
https://doi.org/10.1007/978-3-030-31038-7_8

that will be easy to link together. In effect we hope to use the results of the Views paper as a conceptual architecture to organise our work.

Work we did developing a denotational semantics called "Unifying Theory of Concurrent Programs" (UTCP), for the Views command language, using near-homogeneous relations [6], exposed the need for well-defined semantic building blocks, with very well defined properties. Validating the theory as it was developed required a lot of test calculations, to uncover its final form, to such an extent that a rapid "prototyping" calculator was developed to assist in this endeavour [5]. This calculator depended crucially on having well-defined laws and algebras for the semantic building blocks. We use the adjective "inner" to refer to these, and the term "outer" applies to the top-level laws and algebras of the language that is under study.

Most investigations into the relationship between algebraic, and denotational or operational semantics for a language focus on how they relate at the top-level (e.g., [25, 26]). In this paper, we focus mainly on those small inner algebras that are very helpful in producing a denotational semantics, against which an (outer) algebraic semantics may be compared.

We next present background material in Sect. 2 on the Views command language, and other approaches to the semantics of shared-state concurrency. In Sect. 3 we give a high level overview of the architecture of UTCP, explaining why we have the inner algebras that we do. In Sect. 4 we explore the inner algebra and laws of UTCP, which we then follow up on in Sect. 5 with discussion of the state of the outer theory. In Sect. 6 we look at related work, with an emphasis on those in which there is clear evidence of these inner algebras or laws. Finally, we conclude (Sect. 7).

2   Background

2.1   View Command Language

The baseline command language from the Views paper assumes an abstract notion of shared state s, and a notion of atomic actions a that non-deterministically modify s. The language syntax then takes atomic commands augmented with skip as a building block and introduces operators for sequencing (; ;), choice (+), parallel composition (∥) and iteration (*), where all choices are non-deterministic [9]:

$$C ::= \langle a \rangle \mid \text{skip} \mid C \mathbin{;;} C \mid C + C \mid C \parallel C \mid C^*$$

An operational semantics is then defined based on the notion of interleaving of atomic actions. Our notation differs slightly from that in [9] in that we write "$\langle a \rangle$" and "$;;$" instead of "a" and "$;$" respectively, for reasons explained in Sect. 3.

2.2   Denotational Semantics

The UTP theory of concurrent programs (UTCP), whose algebras we discuss here, gives a denotational semantics to the command language above [6]. Denotational semantics of shared variable concurrency are not new, with notable work in this area having been done by de Boer [1] and Brookes [2]. This resulted in semantics based on the notion of transition traces (TT), which are sequences of state-pairs. A state pair (s_a, s_b) denotes the occurrence of some atomic action that transformed state s_a into state s_b. A transition trace is a sequence of such pairs, with no requirement for the second state of one pair to match the first state of the next. So, if s_i for $i \in 1 \ldots 4$ denotes four different states, then $\langle (s_1, s_2), (s_3, s_4) \rangle$ is valid. It states that the command to which it refers first altered s_1 to s_2, and then something else in the environment ran, changing the state to s_3 along the way, so that when the command resumed to perform its second atomic action, it saw s_3, which it duly converted to s_4. This is basically how the interference of the environment is modelled. The denotational semantics of a command is a set of such traces, with three important healthiness conditions that describe closures:

- *Stuttering:* for any trace $\langle \ldots, (s_1, s_2), (s_3, s_4), \ldots \rangle$
  there is also a trace $\langle \ldots, (s_1, s_2), (s, s)(s_3, s_4), \ldots \rangle$ for arbitrary s.
- *Mumbling:* for any trace $\langle \ldots, (s_1, s_2), (s_3, s_4), \ldots \rangle$ where $s_2 = s_3$
  then there is also a trace $\langle \ldots, (s_1, s_4), \ldots \rangle$.
- *Interference:* for any trace $\langle \ldots, (s_1, s_2), (s_3, s_4), \ldots \rangle$
  there is also a trace $\langle \ldots, (s_1, s_2), (s_5, s_6)(s_3, s_4), \ldots \rangle$ for arbitrary s_5 and s_6.

So the semantics is a set of transition traces closed by adding every possible stuttering action, all possible mumblings, and all possible interference by any possible environment. With the exception of the semantics of a single atomic action, these are all infinitely large sets of traces. These closed sets are fine, when their purpose is to prove that the desired algebraic laws hold for the language under consideration. There is a UTP treatment of Views by van Staden [23] in which he makes use of finite transition traces within an operational calculus.

Another interesting approach to a denotational semantics for shared state concurrency was that reported by Lamport [18]. It is based on the use of temporal logic along with five key ideas, some his, some from others:

1. Being able to identify "who" performs an action.
2. Statement assertions true only if true of every program containing that statement.
3. Being able to transform an assertion about a statement into one about a larger statement that contains it.
4. Defining relations between control points as aliasing relations among variables.
5. Allowing *stuttering* actions, to facilitate decomposing atomic actions.

The UTCP theory we describe here was developed before Lamport's work was discovered, but it is interesting to note how our semantics required the rediscovery of concepts analogous to some of the ideas above.

2.3   UTP Action Semantics

A UTP semantics for parallel programming (UTPP) was developed by Woodcock and Hughes [24], that considered a language that required all atomic actions, and some instances of composite commands to have unique labels. They mapped the language into an action system, where every atomic command became a guarded action, with the guard asserting that the action's label was "enabled", this being modelled by it being present in a global label-set ls. Each guarded action, when enabled, would perform its atomic state-change, remove its enabling label from ls and then add in labels to enable other guarded actions. In effect, the global label set was used to manage flow of control. This theory has the following observations:

$$s, s' : State \tag{1}$$

$$ls, ls' : \mathcal{P}\, Lbl \tag{2}$$

An atomic action, described as a relation $a : State \leftrightarrow State$ with label go, followed by some "after-label" $next$ (say) would exhibit the following behaviour:

$$go \in ls \wedge a \wedge ls' = (ls \setminus \{go\}) \cup \{next\} \tag{3}$$

An action-system is a loop that makes a non-deterministic choice, on each iteration, of one of the currently enabled actions to run. The chosen action will change the state, disable itself, and enable something else.

3   Approach

Our goal for a UTP semantics was to obtain one that was not only compositional (denotational), but was also "local", in the sense that the semantics would only talk about the behaviour of the command under consideration, without being required to also explicitly mention all possible interference. This goal was inspired by the success of separation logic at being able to scale to automatically check very large codebases for pointer errors [21]. A key enabler of that success is that separation logic allows the reasoner to focus on the few pointers actually being manipulated by a program, rather than having to consider (or quantify over) all possible heaps.

We now present a high-level overview of how UTCP is structured, using a simple running example, where a, b and c are arbitrary atomic actions:

$$(\langle a \rangle \;;; \langle b \rangle) \parallel \langle c \rangle \tag{4}$$

The occurrence of action c will be non-deterministically interleaved with those of a and b. Action a will occur before action b. The three possible action sequences (traces) we might observe, assuming no outside interference, are:

$$a; b; c \quad a; c; b \quad c; a; b$$

Here we have represented the sequences using sequential composition (;) which is typically defined in UTP, using O to stand for all observation variables, as:

$$P; Q \; \hat{=} \; \exists O_m \bullet P[O_m/O'] \wedge Q[O_m/O]$$

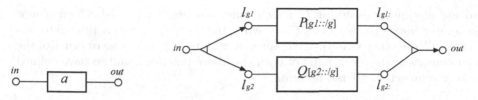

Fig. 1. Atomic action $\langle a \rangle$.   **Fig. 2.** Parallel composition $P \parallel Q$.

This then raises the question of why we use ;; in our command syntax for sequencing. The simple reason is that ;; is not sequential composition as defined in Eq. 5, because that definition not only requires the starting state of Q to be the ending state of P, but it also implies that no interference can occur between the end of P and the start of Q. In our example, an execution of $\langle c \rangle$ can come between $\langle a \rangle$ and $\langle b \rangle$. In effect, in any programming language that admits concurrent threads and shared variables, the semantics of sequencing (;;) is *not* sequential composition as defined above. Instead, it is a form of "loose" sequencing in that its first component must terminate before the second can start, but it places no constraints on how the state might be altered in between.

3.1 Labels and Generators

We provide a semantics based on shared-state s and control-flow label-sets ls, in a fashion similar to Eq. 3. This explains why we use $\langle a \rangle$ to denote the atomic command that performs atomic state-change a, as the latter modifies only s, while the former modifies both s and ls. Unlike UTPP, we do not have explicit labels in our command syntax, but instead we allow the semantic rules to generate appropriate *unique* labels, in a very controlled fashion.

We assume that two label-valued observation variables are associated with every command (atomic or composite), called in and out. The command starts executing when the label that is the value of in is put into the global label set ls. As the command executes, label in will be removed from ls, and eventually, as the command terminates, label out will appear in ls'. The simplest instance of this is an atomic action, represented symbolically in Fig. 1, with a simplified version of its behaviour shown in Eq. 5.

$$in \in ls \wedge a \wedge ls' = (ls \setminus \{in\}) \cup \{out\} \tag{5}$$

The term above is quite complex, so we introduce a the following shorthand, where E and N are label-sets, and a is an atomic action:

$$A(E|a|N) \mathrel{\widehat{=}} E \subseteq ls \wedge a \wedge ls' = (ls \setminus E) \cup N \tag{6}$$

We typically write E or N by listing the labels, without full set-notation, writing $A(in|a|out)$ rather than $A(\{in\}|a|\{out\})$.

In order to give semantics to composite commands, we need to utilise a way to generate labels that guarantees their uniqueness. We require two ways to use any given generator G : Gen: one that generates a label and a new generator (new : $Gen \to Lbl \times Gen$), while the other splits a generator into two ($split$: $Gen \to Gen \times Gen$). We frequently wish to single out one or other of the components of the result pairs of both the above functions, and so have defined a very compact shorthand notation.

$$
\begin{aligned}
G \in Gen ::= {} & g && \text{the ``root'' generator} \\
\mid {} & G_: && \text{generator after label produced from } G \\
\mid {} & G_1 && \text{first generator from split of } G \\
\mid {} & G_2 && \text{second generator from split of } G \\
L \in Lbl ::= {} & \ell_G && \text{label produced by generator } G
\end{aligned}
$$

So, for example, $\ell_{g:2}$ is shorthand for:

$$(fst \circ new \circ snd \circ split \circ snd \circ new)(g)$$

Here Gen is an unusual expression language in that it has only *one* variable g, and three postfix operators, $_:$, $_1$, and $_2$. Even more unusual is the label expression language Lbl, which consists solely of the application of prefix function ℓ_- to a generator, to get the label it generates. Given a generator G we can (i) get a new label and generator: $G \mapsto (\ell_G, G_:)$, or (ii) split a generator: $G \mapsto (G_1, G_2)$.

3.2   Semantics with Generators

The way we use generators in our semantics is to introduce a new observation variable g that denotes a label generator that is available. The intuition behind its use is that a composite construct will use g to generate labels for its own use, and split it and pass the resulting generators into its subcomponents. Passing generator G into sub-component P is simply achieved using substitution: $P[G/g]$. In general, a composite may also need to add some *control-flow* actions, which modify ls, but do not alter s. These perform an identity action (ii) on state:

$$ii \mathrel{\hat=} s' = s \tag{7}$$

Let us now consider sequencing $\langle a \rangle$ and $\langle b \rangle$. In effect, we want to arrange it so that the *out* label of $\langle a \rangle$ is the same as the *in* label of $\langle b \rangle$. We do this by taking generator g, and applying the prefix ℓ operator to obtain label ℓ_g, which then is substituted appropriately. We then take the disjunction of the two modified atomic actions, which treats both actions as being part of a non-deterministic choice, as is done in UTPP [24].

$$\langle a \rangle[\ell_g/out] \lor \langle b \rangle[\ell_g/in] \tag{8}$$

$$= A(in|a|out)[\ell_g/out] \lor A(in|b|out)[\ell_g/in] \tag{9}$$

$$= A(in|a|\ell_g) \lor A(\ell_g|b|out) \tag{10}$$

This is not the full picture as there are healthiness conditions to be applied, and one of them is decidedly non-trivial in its effects. These will be presented and discussed later.

Now, consider putting $\langle c \rangle$ in parallel with $\langle a \rangle \,;;\, \langle b \rangle$. We show the general case for arbitrary commands P and Q in Fig. 2. In effect we split a generator, denoted by g into two (g_1, g_2), generate two labels from each to replace the in and out of P and Q and then add two special control-flow actions. The first replaces the in label for the parallel construct as a whole, by the two generated in-labels (l_{g1}, l_{g2}). The second replaces the two generated out-labels $(l_{g1:}, l_{g2:})$ by the out label of the whole construct. In essence, given P and Q to be put into parallel, we make use [1] of the following disjunction of two control actions and the two components with appropriate label and generator substitutions:

$$A(in|ii|l_{g1}, l_{g2})$$
$$\lor\, P[g_{1::}, l_{g1}, l_{g1:}/g, in, out]$$
$$\lor\, Q[g_{2::}, l_{g2}, l_{g2:}/g, in, out]$$
$$\lor\, A(l_{g1:}, l_{g2:}|ii|out)$$

In our running example, P would be our semantics for $\langle a \rangle \,;;\, \langle b \rangle$ and Q would be $\langle c \rangle$. The latter, after substitution, would appear as $A(l_{g2}|c|l_{g2})$. The former will contain $A(l_{g1}|a|\ell_{g1::})$ and $A(\ell_{g1::}|b|l_{g1:})$, as well as extra components introduced by the healthiness conditions.

3.3   Static and Dynamic Observables

In summary, in addition to observables s, s', ls, ls', we have added in, out, and g. However, this new trio of variables is quite distinct in character, in that they are used within composites to put sub-components into context, by performing substitutions on generators and labels. This contextualisation is *static*, in that it depends on the structure of the program, and it does not change over time. By contrast, observables like s and ls are *dynamic*: they track observables whose values change over time. This distinction is key to making the semantics work.

We now proceed to define sequential composition in our theory in the standard way, provided that we only reference dynamic variables:

$$P; Q \mathrel{\widehat{=}} \exists s_m, ls_m \bullet P[s_m, ls_m/s', ls'] \land Q[s_m, ls_m/s, ls] \tag{11}$$

We can also introduce our notion of Skip (II) which is an identity for sequential composition:

$$II \mathrel{\widehat{=}} s' = s \land ls' = ls \tag{12}$$

No mention is made in either of the above definitions of in, out, or g — these are static, and have no dashed counterparts.

We can now present a complete definition of the alphabet of UTCP predicates:

$$s, s' : State \qquad ls, ls' : \mathcal{P}Lbl \qquad in, out : Lbl \qquad g : Gen$$

---

[1] We have to apply healthiness conditions as well, discussed later.

3.4   Wheels-within-Wheels

Another point to note is that our semantics converts any command into a large disjunction (non-deterministic choice) of atomic and control-flow actions, which precisely correspond to the guarded actions produced in the UTPP semantics [24]. In UTPP, a predicate transformer (run) is applied to the disjunction which initialises the ls and then iterates until/if a distinguished termination label appears in it, generating all possible complete execution traces. This disjunction, produced by both UTPP and UTCP, is static, and in UTPP, run effectively produces the dynamic behaviour. In UTCP, we wanted the predicates generated at every level to capture both static structure, and dynamic behaviour. The intuition was to find a way to "run" at every level in a command program. Each atomic action would be trying to spin continually, awakening when its in label appeared in the label set ls. We need a healthiness condition, to ensure that atomic actions within iterations "stay alive", which basically says the possible behaviour of a command is logically equivalent to making a non-deterministic choice to iterate it zero or more times This healthiness condition has to be applied to the semantics of every command, atomic and composite, leading us to call it "Wheels within Wheels" ($\mathbf{WwW}$). Getting the definition of $\mathbf{WwW}$ right was a major challenge, that drove the development of the rapid-prototype calculator reported in [5]. The key was that we needed to iterate $P \vee II$, which effectively means adding in the possibility of a stuttering step everywhere, leading to the following definition:

$$P^0 \mathrel{\widehat{=}} II \tag{13}$$

$$P^{i+1} \mathrel{\widehat{=}} P \mathbin{;} P^i \tag{14}$$

$$\mathbf{WwW}(P) \mathrel{\widehat{=}} \bigvee_{i \in \mathbb{N}} P^i \tag{15}$$

One key observation here is that with UTCP we need fairly ubiquitous stuttering, just as found in the other compositional theories discussed earlier. Another somewhat striking observation is that we produce a potentially infinite disjunction of P sequentially composed with itself multiple times! This presents quite a challenge for the use of this semantics, and was a key motivation for the UTP Calculator development [5], but it is key to making things work. Given iteration-free programs, the number of iterations actually required is bounded.

Importantly, for a healthiness condition, $\mathbf{WwW}$ is indeed both idempotent and monotonic:

$$P \sqsubseteq Q \implies \mathbf{WwW}(P) \sqsubseteq \mathbf{WwW}(Q) \tag{16}$$

$$\mathbf{WwW}(\mathbf{WwW}(P)) = \mathbf{WwW}(P) \tag{17}$$

3.5   Label Healthiness

We also need some healthiness conditions on labels, generators, and the global label-set ls. One, "Disjoint labels" ($\mathbf{DL}$), requires that $in \neq out$, and neither in nor out appear in $labs(g)$. We introduce more shorthand, using $\{L_1|L_2| \dots |L_n\}$

for $L_1 \uplus L_2 \uplus \cdots \uplus L_n$, and using G as a shorthand for $labs(g)$ in a context where a label-set is expected rather than a generator. This allows us to write the disjoint label condition as $\{in|g|out\}$. This is a *static* invariant that needs to be satisfied by all healthy P:

$$\mathbf{DL}(P) \stackrel{\wedge}{=} P \wedge \{in|g|out\} \tag{18}$$

Another, "Label Exclusivity" (**LE**), requires that labels in, out can never occur together in ls, and also never when any member of $labs(g)$ is present. Again, we introduce a shorthand $[L_1|L_2] \ldots |L_n]$ which asserts for any two different L_i and L_j, that $L_i \cap ls \neq \emptyset \implies L_j \cap ls = \emptyset$. We also use $[L_1|L_2] \ldots |L_n]'$ to denote the above assertion with ls replaced by ls' throughout. This is now a *dynamic* invariant that needs to be satisfied by all healthy P:

$$\mathbf{LE}(P) \stackrel{\wedge}{=} P \wedge [in|g|out] \wedge [in|g|out]' \tag{19}$$

Both **DL** and **LE** are clearly idempotent and monotonic w.r.t refinement.

We can now define a top-level healthiness condition called **W**:

$$\mathbf{W}(P) \stackrel{\wedge}{=} \mathbf{DL}(\mathbf{LE}(\mathbf{WwW}(P))) \tag{20}$$

So our full definitions of $\langle a \rangle$ and $P \parallel Q$, and the other constructs, for completeness, can now be shown in Fig. 3.

$$\langle a \rangle \stackrel{\wedge}{=} \mathbf{W}(\ A(in|a|out)\ ) \tag{21}$$

$$\mathtt{skip} \stackrel{\wedge}{=} \langle ii \rangle \tag{22}$$

$$P \mathbin{;;} Q \stackrel{\wedge}{=} \mathbf{W}(P[g_{:1}, \ell_g/g, out] \vee Q[g_{:2}, \ell_g/g, in]) \tag{23}$$

$$\begin{aligned} P \parallel Q \stackrel{\wedge}{=} \mathbf{W}(\ &A(in|ii|\ell_{g1}, \ell_{g2}) \vee \\ &P[g_{1::}, \ell_{g1}, \ell_{g1:}/g, in, out] \vee \\ &Q[g_{2::}, \ell_{g2}, \ell_{g2:}/g, in, out] \vee \\ &A(\ell_{g1:}, \ell_{g2:}|ii|out)\ ) \end{aligned} \tag{24}$$

$$\begin{aligned} P + Q \stackrel{\wedge}{=} \mathbf{W}(\ &A(in|ii|\ell_{g1}) \vee A(in|ii|\ell_{g2}) \vee \\ &P[g_{1::}/g] \vee Q[g_{2::}/g] \vee \\ &A(\ell_{g1:}|ii|out) \vee A(\ell_{g2:}|ii|out)\ ) \end{aligned} \tag{25}$$

$$\begin{aligned} P^* \stackrel{\wedge}{=} \mathbf{W}(\ &A(in|ii|\ell_g) \vee \\ &A(\ell_g|ii|\ell_{g:}) \vee \\ &A(\ell_g|ii|out) \vee \\ &P[g_{::}, \ell_{g:}, \ell_g/g, in, out]\ ) \end{aligned} \tag{26}$$

Fig. 3. Command semantics in UTCP

In the following sections, we look at in more detail to uncover the laws and algebras that underpin the semantics just described.

4   Inner Algebras

Here we present some of the algebras and laws that characterise the underlying semantic domains of UTCP.

4.1   Labels and Generators

We need to be sure that however we implement, or model, label generation, that we are sure that unique labels are produced. We can give a minimal specification by positing a function *labs* that takes a generator as input, and returns the set of all labels it could possibly generate, and then requiring that: (i) any label produced from a call to *new* can never occur in the modified generator returned by that call, and (ii) the two generators produced by *split* have disjoint label-sets:

$$labs : Gen \rightarrow \mathcal{P}\, Lbl \tag{27}$$

$$\ell_G \notin labs(G_:) \tag{28}$$

$$labs(G_1) \cap labs(G_2) = \emptyset \tag{29}$$

There is a simple way to model/interpret such generators and labels that automatically satisfies the above requirements, plus the following stricter one ($\uplus$ is disjoint union):

$$labs(G) = \{\ell_G\} \uplus labs(G_:) \uplus labs(G_1) \uplus labs(G_2) \tag{30}$$

We simply take labels and generators to be generator expressions themselves, interpreted as strings starting with 'g' and followed by zero or more ':', '1', and '2'. The ℓ operator returns the generator string as the label. The :, 1, and 2 operators append ':', '1', and '2' respectively to the end of the generator string[2]. The advantages of this are two-fold. First, it's simple to describe (and implement, if needed), compared to trying to produce labels as natural numbers (say), that satisfy the requirements above. In particular, there is no need to have a central pool of already generated labels that can be accessed by all the generators that result from *new* and *splits*. Secondly, this interpretation of labels as sequences of symbols that basically record how they were generated from some 'root' generator g, gives us a very easy way to support some of the ideas of Lamport [18], (notably 1, 3, and 4, on p3). Given a top-level command P that mentions atomic action $A(\ell_G|a|\ell_{G:})$, then G, as a string, identifies the path from the top-level down to that atomic action, so answering the "who" question (idea 1).

Performing a substitution of G_a for g in G_b ($G_b[G_a/g]$) is equivalent to generator string concatenation, resulting in G_{ab}. Given two generator strings G_p and G_q associated with commands P and Q say, we can answer questions such as: (i) is P a sub-component of Q (G_q a prefix of G_p)? or (ii) do P and Q have a parent component in common, other than the top-level (G_p and G_q have a common

---

[2] A form of Herbrand interpretation!.

prefix)? A key point to understand about the semantics is that the way substitutions for g, in and out are used maintains this simple relationship between components and sub-components, which makes it easy to facilitate Lamport's idea 3.

4.2   Ground Expressions

The distinction between the dynamic observables (s, s', ls, ls') and the static ones (in, out, g) is crucial. In particular, the relationship between substitution and sequential composition is key. Consider applying a substitution σ to the results of a sequential composition of P and Q. On what circumstances should substitution distribute in through such a composition?

$$(P; Q)\sigma =_? P\sigma; Q\sigma$$

If σ involves dynamic observations, then this should clearly not hold. However, all the substitutions in our semantics are used to modify labels in a systematic way down through a sub-component. So if σ only involves static observations, then we do want this distributive property.

Another is issue is our use of the healthiness condition $\mathbf{WwW}$ in our semantics. This effectively replaces P by a nondeterministic iteration of $P \vee II$, which performs an arbitrary number (zero or more) of sequential compositions. We want certain predicates, and substitutions, to distribute through $\mathbf{WwW}$.

To this end, we first define the notion of a *ground term*, as one that only refers to the static observables g, in, and out. A notable example from our semantics is the invariant $\{in|g|out\}$ associated with the $\mathbf{DL}$ healthiness condition.

Ground predicates K distribute freely through semantic sequential composition:

$$K \wedge (P; Q) = (K \wedge P); Q \tag{31}$$

$$= P; (K \wedge Q) \tag{32}$$

$$= (K \wedge P); (K \wedge Q) \tag{33}$$

These are all an easy consequence of the way in which we defined sequential composition to only involve the dynamic observables. We also note that sequential composition is idempotent on ground predicates, which clearly indicates that the observables in, out, and g, are truly unchanging.

$$K; K = K \tag{34}$$

Finally, we can show that ground predicates K freely distribute in and out of $\mathbf{WwW}$:

$$K \wedge \mathbf{WwW}(P) = \mathbf{WwW}(K \wedge P) \tag{35}$$

A *ground substitution* γ is one where the target variables are static, and all the replacement terms (G, I, O) are ground. Such a substitution will have the following most general form:

$$\gamma = [G, I, O/g, in, out] \tag{36}$$

A key property of ground substitutions is that they distribute through sequential composition (and also have no effect on II):

$$(P \, ; Q)\gamma = P\gamma \, ; Q\gamma \tag{37}$$

$$II\gamma = II \tag{38}$$

They also distribute into the label-set healthiness conditions.

$$\{L_1| \dots |L_n\}\gamma = \{L_1\gamma| \dots |L_n\gamma\} \tag{39}$$

$$[L_1| \dots |L_n]\gamma = [L_1\gamma| \dots |L_n\gamma] \tag{40}$$

We also have the result that the composition of two ground substitutions is itself ground:

$$[G_1, I_1, O_1/g, in, out]\gamma_2 = [G_1\gamma_2, I_1\gamma_2, O_1\gamma_2/g, in, out] \tag{41}$$

4.3   Sound Substitutions

Consider the ground substitution $[g, \ell_g, \ell_g/g, in, out]$ applied to the label disjointedness assertion $\{in|g|out\}$. We obtain the result $\{\ell_g|g|\ell_g\}$, which violates the **DL** healthiness condition. To prevent this we need the notion of a sound substitution ς, which is a ground substitution where the three replacement expressions themselves collectively satisfy **DL**:

$$\varsigma = [G, I, O/g, in, out] \ \textbf{where} \ \{I|G|O\} \tag{42}$$

We note that soundness is also preserved by substitution composition, and also that every substitution used in the semantics is sound.

The disjoint-label healthiness condition predicate is ground, so **DL** distributes through sequential composition

$$\textbf{DL}(P \, ; Q) = \textbf{DL}(P) \, ; \textbf{DL}(Q) \tag{43}$$

The label exclusivity invariant mentions dynamic observables ls and ls', so we only get a weaker form of distributivity:

$$\textbf{LE}(P) \, ; \textbf{LE}(Q) = \textbf{LE}(\textbf{LE}(P) \, ; \textbf{LE}(Q)) \tag{44}$$

$$= \textbf{LE}(P \, ; \textbf{LE}(Q)) \tag{45}$$

4.4   Actions

The core of the UTCP semantics is the notion of a labelled atomic action $A(E|a|N)$. It is enabled when $E \subseteq ls$, and if "chosen" to execute, performs a state change $s \mapsto s'$ that is consistent with the relation a. The semantics of any command reduces to a tree of disjunctions of these, wrapped in the healthiness conditions at every level. The effect of **WwW** is to perform lots of sequential compositions of these with themselves. What is of considerable importance, consequently, is how labelled atomic actions interact with sequential composition.

We start by considering an action composed with itself:

$$A(E|a|N) \; ; \; A(E|a|N) = E \subseteq N \wedge A(E|a^2|N) \tag{46}$$

All atomic actions in the semantics use labelled actions that satisfy **DL**, in which case we have $E \cap N = \emptyset$. For these actions the above self-composition yields **false**. If we consider the full semantics for $\langle a \rangle$:

$$\mathbf{DL}(\mathbf{LE}(\mathbf{WwW}(A(in|a|out))))$$

then the computation of $A(in|a|out)^2$ required by **WwW** is **false**, because we have invariant $\{in|g|out\}$, and so are all the subsequent compositions. So $\mathbf{WwW}(A(in|a|out))$ becomes $II \vee A(in|a|out)$, giving the result:

$$\langle a \rangle = \{in|g|out\} \wedge [in|g|out] \wedge (II \vee A(in|a|out))$$

Unfortunately, our labelled atomic actions are not closed under sequential composition, except under certain conditions, the most notable being when the two actions have no labels in common. In most cases however, we have to introduce an extended form, that differentiates between the enabling labels (E) and those then removed (R). Our basic labelled action is then defined setting $R = E$.

$$X(E|a|R|A) \mathrel{\hat{=}} E \subseteq ls \wedge a \wedge ls' = (ls \setminus R) \cup A \tag{47}$$

$$A(E|a|N) = X(E|a|E|N) \tag{48}$$

We can calculate that the composition of two extended actions is an extended action, provided the label-sets involved satisfy certain conditions that basically ensure that the second action is enabled after the first one runs.

$$X(E_1|a|R_1|A_1); X(E_2|b|R_2|A_2)$$
$$= E_2 \cap (R_1 \setminus A_1) = \emptyset$$
$$\wedge X(E_1 \cup (E_2 \setminus A_1) \mid a \;;_s b \mid R_1 \cup R_2 \mid (A_1 \setminus R_2) \cup A_2) \tag{49}$$

Here we introduce sequential composition restricted to s and s':

$$a \;;_s b \mathrel{\hat{=}} \exists s_m \bullet a[s_m/s'] \wedge b[s_m/s], \tag{50}$$

$$ii \;;_s a = a = a \;;_s ii \tag{51}$$

The result for composing two original atomic actions to produce an extended one is then easily obtained .

$$A(E_1|a|N_1) \; ; \; A(E_2|b|N_2)$$
$$= E_2 \cap (E_1 \backslash N_1) = \emptyset$$
$$\wedge \, X(E_1 \cup (E_2 \backslash N_1) \mid a \; ;_s \, b \mid E_1 \cup E_2 \mid (N_1 \backslash E_2) \cup N_2) \tag{52}$$

We finish actions by noting that ground substitutions distribute into their labels:

$$A(E|a|N)\gamma = A(E\gamma|a|N\gamma) \qquad \text{«·A-gamma-subs·»}$$
$$X(E|a|R|A)\gamma = X(E\gamma|a|R\gamma|A\gamma)$$

4.5   Invariants

Healthiness conditions **DL** and **LE** introduce invariants such as $\{in|g|out\}$, $[in|g|out]$ and/or $[in|g|out]'$. The execution of an atomic action cannot alter the truth of the first one, but it can effect the other two. We require atomic action commands, including control-flow actions, to preserve the **LE** invariant. For basic atomic actions, this is straightforward, and we can show it holds under any sound substitution ς, which covers all the uses of atomic actions as sub-components of composite commands.

$$\{in|g|out\}\varsigma \wedge [in|g|out]\varsigma \wedge A(in|a|out)\varsigma \implies [in|g|out]'\varsigma \tag{53}$$

Finally, given extended actions X_1 and X_2, we have a law about how they interact with law invariants (I_1 and I_2) :

$$(I_1 \wedge X_1) \; ; \; (I_2 \wedge X_2) = I_1 \wedge I_2[(ls \backslash R_1) \cup A_1/ls] \wedge (X_1 \; ; \; X_2) \tag{54}$$

$$= I_{12} \wedge (X_1 \; ; \; X_2) \tag{55}$$

$$\textbf{where } I_{ij} = I_i \wedge I_j[(ls \backslash R_i) \cup A_i/ls] \tag{56}$$

Given that invariants are preserved, as a result of careful theory construction, we might ask if they can be dropped to simplify matters, especially **LE** whose distributivity is limited. Unfortunately, we can't omit them, as they are very useful when doing calculations. It turns out that every instance of $X(\ldots)$ that is left over, can be converted back into an equivalent instance of $A(\ldots)$, because the invariant gives extra information to allow this simplification. A canonical example of this is $X(L_1|a|L_1, L_2|L_3)$ given invariant $[L_1|L_2|\ldots]$. The action is enabled if $L_1 \subseteq ls$, removes L_1 and L_2 from ls, and adds in L_3. However the invariant forbids L_2 from being in ls when this action is enabled, so the removal of L_2 is superfluous. So when enabled, the above action is equivalent to $X(L_1|a|L_1|L_3)$, which is the same as $A(L_1|a|L_3)$, by Eq. 48.

5   Outer Algebras

The notion of Concurrent Kleene Algebra (CKA) $(A, +, *, ;, ^{\circledast}, ^{\circlearrowleft}, 0, 1)$ is being put forward as a baseline for the semantics of all programming languages [16], and is defined as a bi-Kleene algebra over concurrent monoid $(A, *, ;, 1, \leq)$. In that paper, a following rough correspondence between CKA operators and those of CSP are given: $+$ is non-deterministic choice, ; is sequential composition, $*$ is some form of parallelism, $\leq$ is refinement, 1 is SKIP, 0 is miracle, and the circled operators are iterated parallel and sequential composition.

The command language does not have all of those operators, and so we cannot claim that its semantics forms a CKA. If we define

$$\texttt{skip} \mathrel{\widehat{=}} \langle ii \rangle \tag{57}$$

$$P \sqsubseteq Q \mathrel{\widehat{=}} P = P + Q \tag{58}$$

then we can posit the following laws:

$$\texttt{skip} \mathbin{;;} P = P$$
$$P \mathbin{;;} \texttt{skip} = P$$
$$P \mathbin{;;} (Q \mathbin{;;} R) = (P \mathbin{;;} Q) \mathbin{;;} R$$
$$P \parallel Q = Q \parallel P$$
$$P \parallel (Q \parallel R) = (P \parallel Q) \parallel R$$
$$P + P = P$$
$$P + Q = Q + P$$
$$P + (Q + R) = (P + Q) + R$$
$$P + Q \sqsubseteq P$$
$$P \sqsubseteq Q \equiv (P + Q) = Q$$
$$P^* = \texttt{skip} + P \mathbin{;;} P^*$$
$$P_1 \mathbin{;;} (P_2 \parallel P_3) \sqsubseteq (P_1 \mathbin{;;} P_2) \parallel P_3$$
$$P_1 \mathbin{;;} P_2 \sqsubseteq P_1 \parallel P_2$$

Here we discuss how we might proceed to prove that these laws are a consequence of our semantics. The best approach is to explore simple examples of the above laws using atomic actions. We will consider the follow three in order, chosen to reveal key issues we have to tackle.

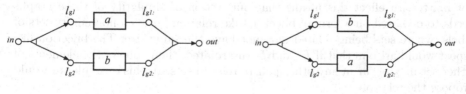

Fig. 4. $\langle a \rangle \parallel \langle b \rangle$ **Fig. 5.** $\langle b \rangle \parallel \langle a \rangle$.

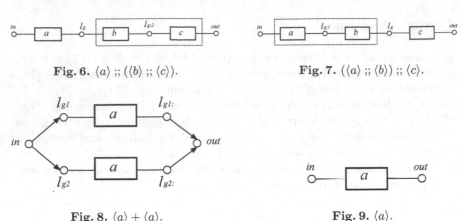

Fig. 6. $\langle a \rangle \;;; (\langle b \rangle \;;; \langle c \rangle)$.

Fig. 7. $(\langle a \rangle \;;; \langle b \rangle) \;;; \langle c \rangle$.

Fig. 8. $\langle a \rangle + \langle a \rangle$.

Fig. 9. $\langle a \rangle$.

$$\langle a \rangle \parallel \langle b \rangle = \langle b \rangle \parallel \langle a \rangle \tag{59}$$

$$\langle a \rangle \;;; (\langle b \rangle \;;; \langle c \rangle) = (\langle a \rangle \;;; \langle b \rangle) \;;; \langle c \rangle \tag{60}$$

$$\langle a \rangle + \langle a \rangle = \langle a \rangle \tag{61}$$

The semantics of parallel can be pictured as per Fig. 2 on p5. The instantiation of this for both sides of the parallel commutativity law example (Eq. 59) are shown in Figs. 4 and 5. We can see that the only difference is that the ℓ_{gNx} labels have been swapped around. This suggests that we should either ignore the particular labels and just look at the structure, of perhaps assume that any bijective mapping of labels has no effect on behaviour.

When we consider the associativity of sequencing (Eq. 60), the two sides are shown in Figs. 6 and 7. Again there is an obvious one-to-one mapping from labels that makes them equivalent.

Things get more complicated when we compare the diagrams (Figs. 8 and 9) for the third example, the idempotence of choice (Eq. 61). This requires more thought: only one of the atomic actions in the lefthand side will run. Unlike the parallel example, where the "production" of both ℓ_{g1} and ℓ_{g2} result from the consumption of in, here we have have two distinct "edges" from label in, so once in is in ls, both edges are enabled, but only one is chosen non-deterministically, so either ℓ_{g1} or ℓ_{g2} are enabled, but not both. As in will be removed from ls, so there is no immediate chance of the other option being enabled. What we have to realise is that control-flow actions are important but from an external observer's perspective, as long as they are well-behaved, the precise details do not matter. In effect this means that our notion of similarity of these graphs needs to be based on a form of bijection-like relation between non-empty sets of labels, where sets being related may not have the same size. The bijection-like aspect would arise in that if L_1 and L_2 are related, then none of the elements of either set may occur in any other pair of related sets. In this example we would propose the relation:

$$\{(\{in, \ell_{g1}, \ell_{g2}\}, \{in\}), (\{\ell_{g1:}, \ell_{g2:}, out\}, \{out\})\}$$

The upshot of all of this, is that one possible plan to prove the laws has two steps: the first is to show that the laws always induce pairs of graphs like the above related by some relational bijection. The second is to show that these graphs induce the behaviour that results from calculating with the semantics.

There is another alternative to be considered: the UTCP semantics define predicates, but their structure is very graph-like. We can view the labelled nodes as vertices of the graph, and labelled actions as action-labelled edges connecting a set of "input" vertices to a set of "output" vertices. We can imagine a vertex coloured black or white to indicate if its label is present in ls. With this graph interpretation of the denotational semantics (predicate) rules, we should be able to produce an operational semantics. This may provide an alternative route to proving the laws.

6   Related Work

The most obvious use of inner algebras in UTP can be found in those theories, usually of concurrency, that make use of trace observations. We can find these in the UTP book, for example the definition of the trace merge operator [15, Definition 8.19, p. 203]. We also see a variant of it used in *Circus* [19]. It is particulary notable in any work adding time to traces, such as *Circus*-time [22], "slotted"-*Circus* [7,13,14], and recent work on trace algebras being used for hybrid semantics [10]. A particularly interesting example of inner algebras and laws, comes from work mechanising UTP in Isabelle/HOL [12]. In this, the traditional pre/post divide becomes a pre/peri/post divide, in which the behaviours when waiting for an external event are captured as distinct peri-conditions [11]. Certain idioms commonly used when defining pre-, peri-, and post-conditions are abstracted out and shown to obey useful laws. These prove to be very valuable for high speed automated proof.

7   Conclusions

While describing work to develop and validate a denotational semantics for the command language in UTP, we have discovered the value of algebra as a tool to help develop such theories. This has been driven by the need to manage the complexity inherent in the underlying semantic domains. It is particularly helpful when the area has conceptual difficulties, and you need healthiness conditions, such as **WwW**, that are highly counter-intuitive. The need for some automation to help assess emerging theories drove the development of the "UTP calculator" described in [5]. What became very clear from that work is that good algebra design leads to very effective and fast calculation, with a lot of scope for automation. The calculator can have been considered a stop-gap measure, but has inspired a complete re-design of the Saoithín/UTP2 theorem prover, originally described in [3,4]. This new version, being developed at https://github.com/andrewbutterfield/reasonEq will support both proof and calculation, with scope for considerable automation.

References

1. de Boer, F.S., Kok, J.N., Palamidessi, C., Rutten, J.J.M.M.: The failure of failures in a paradigm for asynchronous communication. In: Baeten, J.C.M., Groote, J.F. (eds.) CONCUR 1991. LNCS, vol. 527, pp. 111–126. Springer, Heidelberg (1991). https://doi.org/10.1007/3-540-54430-5_84
2. Brookes, S.D.: Full abstraction for a shared-variable parallel language. Inf. Comput. **127**(2), 145–163 (1996). https://doi.org/10.1006/inco.1996.0056
3. Butterfield, A.: Saoithín: a theorem prover for UTP. In: Qin, S. (ed.) UTP 2010. LNCS, vol. 6445, pp. 137–156. Springer, Heidelberg (2010). https://doi.org/10.1007/978-3-642-16690-7_6
4. Butterfield, A.: The logic of $U \cdot (TP)^2$. In: Wolff, B., Gaudel, M.-C., Feliachi, A. (eds.) UTP 2012. LNCS, vol. 7681, pp. 124–143. Springer, Heidelberg (2013). https://doi.org/10.1007/978-3-642-35705-3_6
5. Butterfield, A.: UTPCalc—a calculator for UTP predicates. In: Bowen, J.P., Zhu, H. (eds.) UTP 2016. LNCS, vol. 10134, pp. 197–216. Springer, Cham (2017). https://doi.org/10.1007/978-3-319-52228-9_10
6. Butterfield, A.: UTCP: compositional semantics for shared-variable concurrency. In: Cavalheiro, S., Fiadeiro, J. (eds.) SBMF 2017. LNCS, vol. 10623, pp. 253–270. Springer, Cham (2017). https://doi.org/10.1007/978-3-319-70848-5_16
7. Butterfield, A., Sherif, A., Woodcock, J.: Slotted-circus. In: Davies, J., Gibbons, J. (eds.) IFM 2007. LNCS, vol. 4591, pp. 75–97. Springer, Heidelberg (2007). https://doi.org/10.1007/978-3-540-73210-5_5
8. Calcagno, C., O'Hearn, P.W., Yang, H.: Local action and abstract separation logic. In: Proceedings of 22nd IEEE Symposium on Logic in Computer Science, LICS 2007, Wroclaw, Poland, 10–12 July 2007, pp. 366–378. IEEE Computer Society (2007)
9. Dinsdale-Young, T., Birkedal, L., Gardner, P., Parkinson, M.J., Yang, H.: Views: compositional reasoning for concurrent programs. In: Giacobazzi, R., Cousot, R. (eds.) The 40th Annual ACM SIGPLAN-SIGACT Symposium on Principles of Programming Languages, POPL 2013, Rome, Italy, 23–25 January 2013, pp. 287–300. ACM (2013). https://doi.org/10.1145/2480359.2429104
10. Foster, S., Cavalcanti, A., Woodcock, J., Zeyda, F.: Unifying theories of time with generalised reactive processes. Inf. Process. Lett. **135**, 47–52 (2018). https://doi.org/10.1016/j.ipl.2018.02.017
11. Foster, S., Ye, K., Cavalcanti, A., Woodcock, J.: Calculational verification of reactive programs with reactive relations and Kleene algebra. In: Desharnais, J., Guttmann, W., Joosten, S. (eds.) RAMiCS 2018. LNCS, vol. 11194, pp. 205–224. Springer, Cham (2018). https://doi.org/10.1007/978-3-030-02149-8_13
12. Foster, S., Zeyda, F., Woodcock, J.: Isabelle/UTP: a mechanised theory engineering framework. In: Naumann, D. (ed.) UTP 2014. LNCS, vol. 8963, pp. 21–41. Springer, Cham (2015). https://doi.org/10.1007/978-3-319-14806-9_2
13. Gancarski, P., Butterfield, A.: The denotational semantics of slotted-Circus. In: Cavalcanti, A., Dams, D.R. (eds.) FM 2009. LNCS, vol. 5850, pp. 451–466. Springer, Heidelberg (2009). https://doi.org/10.1007/978-3-642-05089-3_29
14. Gancarski, P., Butterfield, A.: Prioritized slotted-Circus. In: Cavalcanti, A., Deharbe, D., Gaudel, M.-C., Woodcock, J. (eds.) ICTAC 2010. LNCS, vol. 6255, pp. 91–105. Springer, Heidelberg (2010). https://doi.org/10.1007/978-3-642-14808-8_7

15. Hoare, C.A.R., He, J.: Unifying Theories of Programming. Prentice-Hall, Upper Saddle River (1998)
16. Hoare, C.A.R., Möller, B., Struth, G., Wehrman, I.: Concurrent Kleene algebra and its foundations. J. Logic Algebraic Program. **80**(6), 266–296 (2011). https://doi.org/10.1016/j.jlap.2011.04.005. http://www.sciencedirect.com/science/article/pii/S1567832611000166. Relations and Kleene Algebras in Computer Science
17. Jones, C.B.: Tentative steps toward a development method for interfering programs. ACM Trans. Program. Lang. Syst. **5**(4), 596–619 (1983). https://doi.org/10.1145/69575.69577
18. Lamport, L.: An axiomatic semantics of concurrent programming languages. In: Apt, K.R. (ed.) Logics and Models of Concurrent Systems. NATO ASI Series (Series F: Computer and Systems Sciences), vol. 13, pp. 77–122. Springer, Heidelberg (1985). https://doi.org/10.1007/978-3-642-82453-1_4
19. Oliveira, M., Cavalcanti, A., Woodcock, J.: A UTP semantics for Circus. Formal Aspects Comput. **21**(1–2), 3–32 (2009). https://doi.org/10.1007/s00165-007-0052-5
20. Owicki, S.S., Gries, D.: An axiomatic proof technique for parallel programs I. Acta Inf. **6**, 319–340 (1976). https://doi.org/10.1007/BF00268134
21. Pym, D., Spring, J.M., O'Hearn, P.: Why separation logic works. Philos. Technol. (2018). https://doi.org/10.1007/s13347-018-0312-8
22. Sherif, A., Jifeng, H.: Towards a time model for *Circus*. In: George, C., Miao, H. (eds.) ICFEM 2002. LNCS, vol. 2495, pp. 613–624. Springer, Heidelberg (2002). https://doi.org/10.1007/3-540-36103-0_62
23. Staden, S.: Constructing the views framework. In: Naumann, D. (ed.) UTP 2014. LNCS, vol. 8963, pp. 62–83. Springer, Cham (2015). https://doi.org/10.1007/978-3-319-14806-9_4
24. Woodcock, J., Hughes, A.: Unifying theories of parallel programming. In: George, C., Miao, H. (eds.) ICFEM 2002. LNCS, vol. 2495, pp. 24–37. Springer, Heidelberg (2002). https://doi.org/10.1007/3-540-36103-0_5
25. Zhu, H., He, J., Qin, S., Brooke, P.J.: Denotational semantics and its algebraic derivation for an event-driven system-level language. Formal Aspects Comput. **27**(1), 133–166 (2015). https://doi.org/10.1007/s00165-014-0309-8
26. Zhu, H., Yang, F., He, J.: Generating denotational semantics from algebraic semantics for event-driven system-level language. In: Qin, S. (ed.) UTP 2010. LNCS, vol. 6445, pp. 286–308. Springer, Heidelberg (2010). https://doi.org/10.1007/978-3-642-16690-7_15

Developing an Algebra for
Rely/Guarantee Concurrency:
Design Decisions and Challenges

Ian J. Hayes(✉) [iD] and Larissa A. Meinicke [iD]

The University of Queensland, Brisbane, Queensland, Australia
Ian.Hayes@uq.edu.au

Abstract. An algebra for rely/guarantee concurrency has been constructed via a hierarchy of algebraic theories starting from basic theories like lattices through to theories of synchronous behaviour of atomic steps and a theory to support localisation. The algebra is supported by a model based on Aczel traces. We examine the role of these theories in developing a mechanised theory for deriving concurrent programs and outline some of the challenges remaining.

Keywords: Concurrency · Rely/guarantee · Program algebra ·
Synchronous parallel · Localisation

1 Introduction

This paper overviews the design decisions made in developing a refinement algebra for concurrent programs and then outlines some issues that remain. The concurrent refinement algebra that we have developed supports the rely/guarantee style of reasoning about concurrent programs [24–26] but we would like to think it is more general than that. Our design principles are to

- focus on *algebraic properties* in order to facilitate reasoning and theory reuse,
- use a small number of *primitive operators and commands* from which more complex constructs are built,
- utilise the principle of *separation of concerns* whenever possible, and
- provide *mechanisations* of the theories (in Isabelle/HOL).

The aim of this paper is to discuss the major role algebra has played in the development of the overall theory. Our approach is

- based on a refinement calculus style in which specifications are first class objects [2,31,33,37] – see Sect. 2,
- follows Kozen's [27] approach in Kleene Algebra with Tests (KAT) in utilising a sub-algebra of tests, in our case instantaneous tests – see Sect. 3,

This research was supported by Discovery Grant DP190102142 from the Australian Research Council (ARC).

P. Ribeiro and A. Sampaio (Eds.): UTP 2019, LNCS 11885, pp. 176–197, 2019.
https://doi.org/10.1007/978-3-030-31038-7_9

- handles assertions, such as preconditions, as well as non-terminating processes (total correctness) and, as in the refinement calculus, a failed assertion allows any behaviour [38] (**abort** in Dijkstra's terminology [9,10]) – see Sect. 4,
- introduces parallel composition via its algebraic properties [13] – see Sect. 5,
- introduces a sub-algebra of atomic steps and utilises a synchronous algebra to handle parallel composition and both weak and strong conjunction [14,17] – see Sect. 6,
- introduces Morgan-style specification commands [32] that are defined in terms of our primitives – see Sect. 7,
- handles rely conditions as assumptions in a manner similar to preconditions [17] – see Sect. 8,
- introduced a *weak conjunction* operator "⋒" [13] in order to allow commands to be conjoined but that respects assumptions by not masking aborting behaviour (i.e. $c \Cap \mathbf{abort} = \mathbf{abort}$) – see Sect. 9,
- handles issues like finite stuttering and mumbling [4,11] by utilising specifications that if they admit an implementation c, admit any program that is equivalent to c modulo finite stuttering and mumbling – see Sect. 10,
- the basic parallel operator has no fairness assumptions but our algebra is rich enough to define a fair parallel operator in terms of a basic parallel operator [15] – see Sect. 11, and
- handles local variables via primitive localisation operators that operate on commands in the same way that existential quantification acts on predicates [12,28] – see Sect. 12.

While Hoare and He's Unifying Theories of Programming [22] provides a unifying (relational) semantic model for theories of programming, we see algebraic theories as unifying in the sense that the one abstract algebraic theory may have multiple instantiations. In that sense our work corresponds more closely to the Laws of Programming approach [21]. Properties that hold in the abstract algebra are usually simpler to prove in the algebra, and can then be reused in each instantiation. This reuse is supported in the Isabelle/HOL theorem prover, making mechanisation simpler.

The remainder of the paper discusses each of the aspects listed above before turning to some remaining challenges for the approach in Sect. 13.

2   Concurrent Refinement Algebra

At the base of our theory we utilise a complete distributive lattice of commands, $\mathcal{C}$, in a manner similar to theories of sequential refinement calculus [3]. This allows one to represent refinement as the lattice order (i.e. $c \sqsubseteq d$ means c is refined (or implemented) by d) and non-determinism as lattice meet operator $c \sqcap d$ that allows the behaviour of either c or d. The command **abort** is the bottom of the lattice and the top of the lattice, $\top$, is the everywhere infeasible command (sometimes called "magic").

The lattice is extended with a sequential composition operator ";" that is associative (1) and has as its identity the null command **nil** (2).[1] The command

---

[1] In algebraic terms it forms a monoid with identity **nil**.

abort is a left annihilator of sequential composition (3) and sequential composition distributes from the right over an arbitrary non-deterministic choice (4) and from the left over a non-empty choice (5). We assume unary operators have the highest precedence, followed by sequential composition and that non-deterministic choice has the lowest precedence; we use explicit parentheses to resolve all other precedence issues.

$$(c_1 \,;\, c_2) \,;\, c_3 = c_1 \,;\, (c_2 \,;\, c_3) \tag{1}$$

$$c \,;\, \mathbf{nil} = c = \mathbf{nil} \,;\, c \tag{2}$$

$$\mathbf{abort} \,;\, c = \mathbf{abort} \tag{3}$$

$$\left(\bigsqcap_{c \in C} c\right) \,;\, d = \bigsqcap_{c \in C} (c \,;\, d) \tag{4}$$

$$c \,;\, \left(\bigsqcap_{d \in D} d\right) = \bigsqcap_{d \in D} (c \,;\, d) \qquad\qquad \text{if } D \neq \varnothing \tag{5}$$

In (5), D is required to be non-empty because $\bigsqcap \varnothing = \top$ and because we require **abort** to be an annihilator from the left (3) as in the sequential refinement calculus,

$$\mathbf{abort} \,;\, \bigsqcap_{d \in \varnothing} d = \mathbf{abort} \neq \top = \bigsqcap_{d \in \varnothing} (\mathbf{abort} \,;\, d).$$

The lattice allows one to define least and greatest fixed points which can be used to define recursive programs and iteration operators, such as finite iteration zero or more times, $c^\star$, possibly infinite iteration zero or more times, c^ω, and infinite iteration c^∞. Following the approach taken by von Wright [38], these operators are defined in terms of least (μ) and greatest (ν) fixed points.

$$c^\star \mathrel{\widehat{=}} \nu x \cdot (c \,;\, x \sqcap \mathbf{nil}) \tag{6}$$

$$c^\omega \mathrel{\widehat{=}} \mu x \cdot (c \,;\, x \sqcap \mathbf{nil}) \tag{7}$$

$$c^\infty \mathrel{\widehat{=}} \mu x \cdot (c \,;\, x) \tag{8}$$

Iteration operators satisfy a range of well known properties, for example, the following.

$$c^\star = c \,;\, c^\star \sqcap \mathbf{nil} = c^\star \,;\, c \sqcap \mathbf{nil} \tag{9}$$

$$c^\omega = c \,;\, c^\omega \sqcap \mathbf{nil} \tag{10}$$

$$(c \sqcap d)^\star = c^\star \,;\, (d \,;\, c^\star)^\star \tag{11}$$

$$(c \sqcap d)^\omega = c^\omega \,;\, (d \,;\, c^\omega)^\omega \tag{12}$$

Our development utilises the algebraic properties of the lattice, fixed points and iteration operators following the work of Cohen [5] and von Wright [38]. While a number of algebraic approaches (e.g. Kleene algebra [27]) only handle partial correctness, in order to handle non-terminating loops and infinite recursion, we take a total correctness approach.

3 Tests

Within his Kleene Algebra with Tests (KAT) Kozen [27] encoded tests as a Boolean sub-algebra of commands. This allowed Kozen to construct conditional and loop constructs in terms of the language primitives,

$$\textbf{if } b \textbf{ then } c \textbf{ else } d \mathrel{\widehat{=}} b \mathbin{;} c \sqcap \overline{b} \mathbin{;} d \tag{13}$$

$$\textbf{while } b \textbf{ do } c \mathrel{\widehat{=}} (b \mathbin{;} c)^\star \mathbin{;} \overline{b} \tag{14}$$

where b is a test and c and d are commands, and $\overline{b}$ is the negation of the test b. The while loop is defined in terms of the iteration operator $c^\star$, which does not model non-terminating loops. This differs the approach used by von Wright [38] that uses $(b \mathbin{;} c)^\omega \mathbin{;} \overline{b}$ instead and hence can handle total correctness. Because our primitives include least fixed points, the loop can also be defined directly as a least fixed point in terms of a conditional command (13).

$$\textbf{while } b \textbf{ do } c \mathrel{\widehat{=}} \mu\, x \cdot (\textbf{if } b \textbf{ then } c \mathbin{;} x \textbf{ else nil}) \tag{15}$$

Tests $(\mathcal{T})$ are a subset of commands that forms a complete Boolean algebra. For modelling programming language conditionals on a state space Σ, a test corresponds to a set of states for which the test succeeds. Hence we define a one-to-one map τ from sets of states to tests, that maps a set of states $p \subseteq \Sigma$ to the test $\tau(p)$. The mapping preserves the Boolean algebra structure of tests and hence is a homomorphism. Because tests form a Boolean algebra, one can reuse the rich theory of lemmas for Boolean algebra provided in the standard theories of Isabelle/HOL to reason about tests. Our tests are instantaneous and hence the true test $\tau(\Sigma)$ is the identity of sequential composition **nil**.

4 Assertions

In Hoare logic [19] preconditions represent assumptions about the initial state of the program. If the precondition does not hold, any behaviour is allowed, and in a total correctness approach that includes non-termination. The refinement calculus includes the command **assert** t that aborts if the test t does not hold but otherwise does nothing [38]. It can be used to encode a precondition at the start of a specification, but as it is a command, it can also be used as an intermediate assertion. Because our language includes the **abort** command, following von Wright [38], **assert** t can be defined as follows, in which $\overline{t}$ is the negation of the test t.

$$\textbf{assert } t \mathrel{\widehat{=}} \overline{t} \mathbin{;} \textbf{abort} \sqcap \textbf{nil} \tag{16}$$

$$= \overline{t} \mathbin{;} \textbf{abort} \sqcap t \tag{17}$$

Because **abort** is a left annihilator, a command such as (**assert** t ; $x := 2$) does not guarantee that the final value of x is 2 when t does not hold initially, even if its implementation terminates. Some approaches are not rich enough to encode such a property. For example, using a relational algebra in which programs are represented as binary relations between states, the most chaotic command is the universal relation, and when that is combined with the relation representing $x := 2$ it does not give the universal relation but a relation that guarantees the final value of x is 2. (If the state space is extended with an additional bottom state representing an aborted state, then an abort command that is a left annihilator can be defined in that extended relational model.)

In the sequential case, non-termination is often conflated with a failing precondition, i.e. the precondition defines the termination set of the program. When dealing with concurrent processes it is best to keep these two concepts separate because a valid behaviour of a command may be non-termination. For example, a command monitoring a nuclear reactor that, when the reactor becomes unsafe, drops its safety rods and terminates, has valid (and desirable) non-terminating behaviour. Our approach distinguishes between **abort** and the non-aborting iteration of a command forever, e.g. **while** *true* **do** *something*.

The assertion command is based on the abstract set of tests. For modelling state-based programs tests are the image of the mapping τ. Hence we define an abbreviation $\{p\}$ for the assertions formed from tests of the form $\tau(p)$.

$$\{p\} \mathrel{\widehat{=}} \textbf{assert}\ \tau(p) \tag{18}$$

5   Parallel

The parallel composition of two commands $c \parallel d$ is introduced via its basic algebraic properties [13]. It is associative (19), commutative (20) and has identity **skip** (21).[2] Note that our algebra does not assume that the identities of sequential composition (**nil**) and parallel composition (**skip**) are the same. The motivation for this is that in Sect. 6 we take a synchronous view of parallel composition rather than an interleaving view; if an interleaving view is taken **nil** and **skip** can be identified. By not identifying them in the basic algebra either interpretation is allowed. Parallel composition is abort strict (22) and distributes over non-empty non-deterministic choices (23).

$$(c_1 \parallel c_2) \parallel c_3 = c_1 \parallel (c_2 \parallel c_3) \tag{19}$$

$$c_1 \parallel c_2 = c_2 \parallel c_1 \tag{20}$$

$$c \parallel \textbf{skip} = c \tag{21}$$

$$c \parallel \textbf{abort} = \textbf{abort} \tag{22}$$

$$\left(\prod_{c \in C} c \right) \parallel d = \prod_{c \in C} (c \parallel d) \qquad \text{if } C \neq \varnothing \tag{23}$$

---

[2] In algebraic terms it forms a commutative monoid with identity **skip**.

When a new operator is introduced one needs to define how it interacts with other operators, for example, (23) shows parallel's interaction with non-deterministic choice. Parallel does not distribute over sequential composition but it does satisfy a weak interchange axiom (24).

$$(c_1 \; ; \; c_2) \parallel (d_1 \; ; \; d_2) \sqsubseteq (c_1 \parallel d_1) \; ; \; (c_2 \parallel d_2) \tag{24}$$

It is common to equate the identities of sequential and parallel composition, i.e. assume **nil** = **skip**. In Concurrent Kleene Algebra (CKA) [23] **nil** and **skip** are equated and hence one can deduce two further laws from (24) by taking c_1, respectively c_2, to be the common identity.

$$c_2 \parallel (d_1 \; ; \; d_2) \sqsubseteq d_1 \; ; \; (c_2 \parallel d_2) \tag{25}$$

$$c_1 \parallel (d_1 \; ; \; d_2) \sqsubseteq (c_1 \parallel d_1) \; ; \; d_2 \tag{26}$$

We do not make such an assumption here and hence do not have these two laws. The reason we do not equate **nil** and **skip** is that we use a synchronous interpretation of the parallel operator (see Sect. 6), which allows us to encode rely conditions as commands in the language (see Sect. 8).

6   Atomic Steps and Synchronous Operators

The approach we have taken to extending the above algebra to support the rely and guarantee conditions of Jones [24–26] is motivated by the approach used by Aczel [1,8] that provides a semantics for rely/guarantee concurrency in terms of traces that distinguish atomic program (direct) and environment (interference) steps. From here on we use the term *step* to mean *atomic step* exclusively. For two processes c and d running in parallel, a program step of c is an environment step of d, a program step of d is an environment step of c, and an environment step of the whole parallel composition $c \parallel d$, i.e. a step of some additional process running in parallel with both c and d, is an environment step of both c and d.

Our initial approach matched Aczel's closely with explicit program and environment steps [14] but we later realised that there was a more abstract atomic steps algebraic structure underlying that. That structure resembles Milner's Synchronous CCS (SCCS) [29,30] and Synchronous Kleene Algebra (SKA) [35], both of which model concurrency by explicit synchronisation of steps of processes and include a "delay" step, called ϵ here, which synchronises with any step, $\mathbf{a}$, of its environment, that is, $\epsilon \parallel \mathbf{a} = \mathbf{a}$. The approach to adding atomic steps, $\mathcal{A}$, is to view them as a subset of commands ($\mathcal{C}$), similar to the way in which tests ($\mathcal{T}$) are embedded as a subset of commands. The algebraic structure we impose on this subset is that, like tests, it forms a complete Boolean algebra, where the Boolean algebra inherits its lattice operators from the command lattice operators and adds a negation operator, $!\mathbf{a}$, on steps. The greatest step command is $\top$ and the least step command is $\boldsymbol{\alpha}$, the step command that allows any atomic step behaviour (see Sect. 6.2 for more details). The negation of step $\mathbf{a}$, $!\mathbf{a}$, can perform any step other than those allowed by $\mathbf{a}$. It differs from the negation operator for tests, $\bar{t}$, because the infeasible command $\top$ is both an atomic step command and a test, but $! \top = \boldsymbol{\alpha}$ while $\overline{\top} = \mathbf{nil}$.

6.1   Abstract Synchronisation Operator

We abstract from the parallel operator and consider the properties of synchronising operators in general. The parallel operator is a synchronising operator, but so is the lattice join operator $\sqcup$, the dual of $\sqcap$, and in Sect. 9 we see yet another synchronising operator ($\text{\textcircled{m}}$). It turns out we can define the behaviour of each of these operators by stating that each is an instance of an abstract synchronising operator "$\otimes$" and then separately defining the behaviour of each operator on pairs of atomic steps [17].

The abstract synchronising operator is associative, commutative and distributes over non-empty non-deterministic choices, as for parallel composition (19), (20), (23). While parallel has identity **skip** (21), the other instances of the synchronisation operator have different identities, and while parallel is abort strict (22), the lattice join ($\sqcup$) is not, so equivalents of (21) and (22) are not assumed for the abstract synchronisation operator ($\otimes$). The additional axioms to handle atomic steps show how its behaviour unfolds one step at a time. If the two processes start with steps **a** and **b**, respectively, and continue with behaviour described by commands c and d, respectively, the atomic steps must first synchronise, $\mathbf{a} \otimes \mathbf{b}$, and then their subsequent behaviours synchronise, $c \otimes d$, i.e. (27). If the **nil** process, that cannot make any step at all, is synchronised with a process that must first perform atomic step **a**, the result is the infeasible program $\top$ (28). Two **nil** processes synchronise to give **nil** (29). For these axioms, **a** and **b** are atomic step commands while c and d are arbitrary commands.

$$\mathbf{a} \,;\, c \otimes \mathbf{b} \,;\, d = (\mathbf{a} \otimes \mathbf{b}) \,;\, (c \otimes d) \tag{27}$$

$$\mathbf{a} \,;\, c \otimes \mathbf{nil} = \top \tag{28}$$

$$\mathbf{nil} \otimes \mathbf{nil} = \mathbf{nil} \tag{29}$$

Axiom (28) ensures that two processes terminate together (via (29)) or both have infinite behaviours that synchronise all steps. If axiom (28) is replaced by the axiom

$$\mathbf{a} \,;\, c \otimes \mathbf{nil} = \mathbf{a} \,;\, c$$

that allows early termination of one component of the synchronisation. Using an early termination form of synchronisation operator for parallel composition does not support rely/guarantee reasoning about concurrency because once a process in a parallel composition terminates early, its rely condition is no longer applicable, and hence the remaining process is free to break the rely condition from that point on.

From axioms (27–29), one can deduce properties about iterations of atomic steps, such as the following for fixed and finite iteration [17], in which c^i represents the fixed iteration of the command c exactly i times, for $i \in \mathbb{N}$.

$$\mathbf{a}^i \otimes \mathbf{b}^i = (\mathbf{a} \otimes \mathbf{b})^i \tag{30}$$

$$\mathbf{a}^\star \otimes \mathbf{b}^\star = (\mathbf{a} \otimes \mathbf{b})^\star \tag{31}$$

$$\mathbf{a}^\star \,;\, c \otimes \mathbf{b}^\star \,;\, d = (\mathbf{a} \otimes \mathbf{b})^\star \,;\, ((c \otimes \mathbf{b}^\star \,;\, d) \sqcap (\mathbf{a}^\star \,;\, c \otimes d)) \tag{32}$$

To handle infinite iteration, $\mathbf{a}^\infty$, we add an additional continuity property that effectively gives us the following axiom.

$$\mathbf{a}^\infty \otimes \mathbf{b}^\infty = (\mathbf{a} \otimes \mathbf{b})^\infty \tag{33}$$

This is needed to deduce the properties equivalent to (30–32) for possibly infinite iterations:

$$\mathbf{a}^\omega \otimes \mathbf{b}^\omega = (\mathbf{a} \otimes \mathbf{b})^\omega \tag{34}$$

$$\mathbf{a}^\omega \; ; \; c \otimes \mathbf{b}^\omega \; ; \; d = (\mathbf{a} \otimes \mathbf{b})^\omega \; ; \; ((c \otimes \mathbf{b}^\omega \; ; \; d) \sqcap (\mathbf{a}^\omega \; ; \; c \otimes d)) \tag{35}$$

One can develop quite a rich algebra from just these basic axioms. Not only is it applicable to the theory we are developing for rely/guarantee concurrency, it also applies to Milner's SCCS [29,30].

6.2   Parallel as a Synchronisation Operator

To instantiate the synchronous algebra for parallel, one needs to define the behaviour of parallel for steps. For rely/guarantee concurrency we need to be more explicit about the structure of atomic steps. The command ϵ allows any step by its environment and its complement in the Boolean atomic step algebra $\pi \mathrel{\widehat{=}} !\,\epsilon$ corresponds to a non-deterministic choice over all possible program steps. The possible effect of a program or environment step on the state space of the program can be represented by a binary relation, r, between states. We therefore define two injective homomorphisms, π and ϵ, to map the Boolean algebra of binary relations on states to the sets of program and environment steps, respectively. For example, for relation r, $\pi(r)$ is the atomic step that can perform any program step from state σ to state σ' provided that $(\sigma, \sigma') \in r$. Because π and ϵ are homomorphisms, they preserve the Boolean algebraic structure of binary relations and hence both sets of program steps and sets of environment steps also form Boolean algebras.

Note that $\boldsymbol{\pi} = \pi(\mathsf{univ})$ and $\boldsymbol{\epsilon} = \epsilon(\mathsf{univ})$, where univ is the universal binary relation on states; $\boldsymbol{\pi}$ and $\boldsymbol{\epsilon}$ are in a bold font to distinguish them from the homomorphisms π and ϵ. Any atomic step command $\mathbf{a}$ can be defined in the form $\pi(r_1) \sqcap \epsilon(r_2)$, for some relations r_1 and r_2. Atomic step commands form a lattice that shares $\top$ as its top element and has $\boldsymbol{\alpha} \mathrel{\widehat{=}} \boldsymbol{\pi} \sqcap \boldsymbol{\epsilon}$ as its least element.

The following axioms define the behaviour of parallel on atomic steps.

$$\pi(r_1) \parallel \epsilon(r_2) = \pi(r_1 \cap r_2) \tag{36}$$

$$\epsilon(r_1) \parallel \epsilon(r_2) = \epsilon(r_1 \cap r_2) \tag{37}$$

$$\pi(r_1) \parallel \pi(r_2) = \top \tag{38}$$

Note that in our algebra two program steps cannot synchronise (i.e. their parallel combination is the infeasible command $\top$), unlike in process algebras like SCCS, which instantiates $\otimes$ in a different manner. Because ϵ is the identity of parallel for a single step, ϵ^ω is the identity of parallel for an arbitrary command, i.e. $\mathbf{skip} = \epsilon^\omega$.

7   Encoding Specification Commands

A Morgan-style specification command, $[p, \ q]$, that terminates in a state satisfying q provided p holds in the initial state [32], can be encoded in our theory, but first we need to define an encoding of termination. The command **term** that only performs a finite number of program steps is defined as follows.

$$\textbf{term} \,\widehat{=}\, \alpha^\star \,;\, \epsilon^\omega \tag{39}$$

Although it allows only a finite number of program steps, it does not preclude the process being preempted by its environment forever and hence may at any stage switch to the behaviour ϵ^ω, which allows only environment steps, possibly forever. Ruling out preemption by the environment can be handled by introducing fairness (see Sect. 11). A command c is *terminating* if it refines **term**, (i.e. if **term** $\sqsubseteq c$).

A command representing a Morgan-style pre-post specification [32] can then be encoded as follows: if the precondition p holds in the initial state, it only performs a finite number of program steps and terminates in a state satisfying q.[3]

$$[p, \ q] \,\widehat{=}\, \{p\} \,;\, \textbf{term} \,;\, \tau(q) \tag{40}$$

8   Encoding Guarantee and Rely Conditions

In the context of rely/guarantee concurrency, a guarantee condition g, a binary relation on states, can be encoded as the command **guar** g that only allows program steps satisfying g but does not restrict environment steps at all.

$$\textbf{guar} \ g \,\widehat{=}\, (\pi(g) \sqcap \epsilon)^\omega \tag{41}$$

It is a refinement to strengthen a guarantee because $\pi(g_1) \sqsubseteq \pi(g_2)$, if $g_2 \subseteq g_1$.

$$\textbf{guar} \ g_1 \sqsubseteq \textbf{guar} \ g_2 \qquad\qquad \text{if } g_2 \subseteq g_1 \tag{42}$$

Section 9 shows how to combine this command with rely conditions and pre/post specifications.

Rely conditions are assumptions about the behaviour of the steps taken by the environment of a process. In order to define a rely condition, we first define a more abstract assumption command that allows any step refining **a** but aborts after any other step, i.e. one refining ! **a**.

$$\textbf{assume a} \,\widehat{=}\, \,!\,\textbf{a} \,;\, \textbf{abort} \sqcap \alpha \tag{43}$$

$$= \,!\,\textbf{a} \,;\, \textbf{abort} \sqcap \textbf{a} \tag{44}$$

Note the similarity to the definition of **assert** (16).

---

[3] For this version q is a set of states; a more general version where q is a binary relation between states can also be defined.

Again one can derive a number of properties of assumptions and, in particular, their iteration, for example, the following lemma.

Lemma 1 (iterate-assumption). *For any atomic step command,* a,

$$(\textbf{assume a})^\omega = \alpha^\omega \; ; (!\,a \, ; \, \textbf{abort} \sqcap \textbf{nil}). \tag{45}$$

Proof. We start by expanding the definition of an assumption (43).

$(!\,a \, ; \, \textbf{abort} \sqcap \alpha)^\omega$

$= \alpha^\omega \; ; (!\,a \, ; \, \textbf{abort})^\omega$ by (12) and **abort** annihilates

$= \alpha^\omega \; ; (!\,a \, ; \, \textbf{abort} \, ; (!\,a \, ; \, \textbf{abort})^\omega \sqcap \textbf{nil})$ unfolding by (10)

$= \alpha^\omega \; ; (!\,a \, ; \, \textbf{abort} \sqcap \textbf{nil})$ as **abort** annihilates

$\square$

A rely condition r, a binary relation on states, can be encoded as a command that aborts if the environment of the command performs a step not satisfying r; it allows any environment steps satisfying r or any program steps.

$$\textbf{rely } r \mathrel{\widehat{=}} (\textbf{assume } (\epsilon(r) \sqcap \boldsymbol{\pi}))^\omega \tag{46}$$

Noting that $!(\epsilon(r) \sqcap \boldsymbol{\pi}) = \epsilon(\overline{r})$, this can be rewritten using (43) as

$$\textbf{rely } r = (\epsilon(\overline{r}) \, ; \, \textbf{abort} \sqcap \alpha)^\omega \tag{47}$$

or applying (45) to (46) gives the following.

$$\textbf{rely } r = \alpha^\omega \; ; (\epsilon(\overline{r}) \, ; \, \textbf{abort} \sqcap \textbf{nil}) \tag{48}$$

Weakening a rely condition is a refinement because if $r_1 \subseteq r_2$ then $\epsilon(\overline{r_1}) \sqsubseteq \epsilon(\overline{r_2})$.

$$\textbf{rely } r_1 \sqsubseteq \textbf{rely } r_2 \qquad\qquad \text{if } r_1 \subseteq r_2 \tag{49}$$

Because relies and guarantees are defined in terms of our atomic step algebra we can prove properties like the following lemma using properties of atomic steps.

Lemma 2 (rely-parallel-guar). *For any binary relation* r,

$$(\textbf{rely } r) = (\textbf{rely } r) \parallel (\textbf{guar } r). \tag{50}$$

Proof. The main property needed is (35).

$(\mathbf{rely}\ r) \parallel (\mathbf{guar}\ r)$

$= \alpha^{\omega} ; (\epsilon(\overline{r}) ; \mathbf{abort} \sqcap \mathbf{nil}) \parallel (\pi(r) \sqcap \epsilon)^{\omega} ; \mathbf{nil}$ by (48) and (41)

$= (\alpha \parallel (\pi(r) \sqcap \epsilon))^{\omega};$ by (35)
$\quad (((\epsilon(\overline{r}) ; \mathbf{abort} \sqcap \mathbf{nil}) \parallel (\pi(r) \sqcap \epsilon)^{\omega}) \sqcap$
$\quad (\alpha^{\omega} ; (\epsilon(\overline{r}) ; \mathbf{abort} \sqcap \mathbf{nil}) \parallel \mathbf{nil}))$

$= \alpha^{\omega};$ by (36–38), (10)
$\quad ((\epsilon(\overline{r}) ; \mathbf{abort} \sqcap \mathbf{nil}) \parallel ((\pi(r) \sqcap \epsilon) ; (\pi(r) \sqcap \epsilon)^{\omega} \sqcap \mathbf{nil}) \sqcap$
$\quad ((\alpha ; \alpha^{\omega} \sqcap \mathbf{nil}) ; (\epsilon(\overline{r}) ; \mathbf{abort} \sqcap \mathbf{nil}) \parallel \mathbf{nil}))$

$= \alpha^{\omega} ; ((\epsilon(\overline{r}) \parallel (\pi(r) \sqcap \epsilon)) ; (\mathbf{abort} \parallel (\pi(r) \sqcap \epsilon)^{\omega}) \sqcap$ by (23) and (27)
$\quad (\epsilon(\overline{r}) ; \mathbf{abort} \parallel \mathbf{nil}) \sqcap$
$\quad (\mathbf{nil} \parallel (\pi(r) \sqcap \epsilon) ; (\pi(r) \sqcap \epsilon)^{\omega}) \sqcap$
$\quad (\mathbf{nil} \parallel \mathbf{nil}) \sqcap$
$\quad (\alpha ; \alpha^{\omega} ; (\epsilon(\overline{r}) ; \mathbf{abort} \sqcap \mathbf{nil}) \parallel \mathbf{nil}))$

$= \alpha^{\omega} ; (\epsilon(\overline{r}) ; \mathbf{abort} \sqcap \top \sqcap \top \sqcap \mathbf{nil} \sqcap \top)$ by (22), (28–29)

$= \mathbf{rely}\ r$ by (48)

$\square$

Property (50) is the foundation of the parallel introduction law for rely/guarantee concurrency Law 3 in Sect. 9.

9   Conjoining Specifications but Respecting Assumptions

Above we have introduced rely and guarantee commands, but we need a way to combine them with specification commands to produce a complete rely/guarantee specification. The lattice join operator $\sqcup$ can be interpreted as a strong form of conjunction (i.e. intersection of trace sets in a trace-based semantic model). It is too strong for use in combining rely, guarantee and specification commands because, although it enforces the *commitments* of all of these, it does not adequately combine their *assumptions*. In both specifications and rely conditions, assumption violations are represented by aborting behaviour, which can be masked by the lattice join, for which $c \sqcup \mathbf{abort} = c$, because $\mathbf{abort}$ is the least element of the lattice. For example, the conjunction of a guarantee with a rely command, $(\mathbf{guar}\ g) \sqcup (\mathbf{rely}\ r)$, simplifies to $(\mathbf{guar}\ g)$: in this example, the assumption specified by the rely command is masked by the guarantee command which can never abort.

What is required is a weak conjunction operator, $\Cap$, that behaves like strong conjunction for non-aborting steps but aborts if either of its arguments aborts, i.e. it is abort strict:

$$c \Cap \mathbf{abort} = \mathbf{abort}. \tag{51}$$

Weak conjunction is a synchronising operator (Sect. 6.1) and hence it is associative and commutative, distributes over non-empty non-deterministic choices, and satisfies the synchronisation axioms (27), (28), (29) and (33). It is also idempotent, i.e.

$$c \Cap c = c \qquad (52)$$

and hence weak conjunction forms a semi-lattice of commands with identity **chaos** (see below). Weak conjunction satisfies the following interchange axioms with sequential and parallel composition.

$$(c_1 \,;\, c_2) \Cap (d_1 \,;\, d_2) \sqsubseteq (c_1 \Cap d_1) \,;\, (c_2 \Cap d_2) \qquad (53)$$

$$(c_1 \parallel c_2) \Cap (d_1 \parallel d_2) \sqsubseteq (c_1 \Cap d_1) \parallel (c_2 \Cap d_2) \qquad (54)$$

We define weak conjunction for pairs of atomic steps as follows.

$$\pi(r_1) \Cap \pi(r_2) = \pi(r_1 \cap r_2) \qquad (55)$$

$$\epsilon(r_1) \Cap \epsilon(r_2) = \epsilon(r_1 \cap r_2) \qquad (56)$$

$$\pi(r_1) \Cap \epsilon(r_2) = \top \qquad (57)$$

These definitions coincide with those for the lattice join ($\sqcup$) because atomic step commands are non-aborting. As a result α ($= \pi \sqcap \epsilon$) is the atomic step identity of weak conjunction, (i.e. $\mathbf{a} \Cap \alpha = \mathbf{a}$ for any atomic step command $\mathbf{a}$), and hence

$$\mathbf{chaos} \,\hat{=}\, \alpha^{\omega} \qquad (58)$$

is the identity of weak conjunction, (i.e. $c \Cap \mathbf{chaos} = c$ for any command c); **chaos** allows any non-aborting behaviour.

Given weak conjunction, a command c satisfies a rely/guarantee specification with precondition p, postcondition q, rely condition r and guarantee condition g, if

$$(\mathbf{rely}\ r) \Cap (\mathbf{guar}\ g) \Cap [p,\ q] \sqsubseteq c. \qquad (59)$$

This specifies that starting from a state satisfying p, in a context in which all environment steps satisfy r, the command c terminates in a state satisfying q and every program step made by c satisfies g. If at any point the environment makes a step not satisfying r, from that point on the specification no longer needs to be satisfied: the program steps no longer need to satisfy the guarantee g; termination is no longer guaranteed; and, even if the program terminates, the final state need not satisfy q. This approach follows that outlined by Jones [6] but differs from Concurrent Kleene Algebra [23], which requires the implementation satisfies the guarantee regardless.

One consequence of our approach is that we are able to give an algebraic proof of the parallel introduction law [17].

Law 3 (parallel-introduction). *If* $c \Cap \mathbf{term} = c$ *and* $d \Cap \mathbf{term} = d$ *and* $c \parallel \mathbf{term} = c$ *and* $d \parallel \mathbf{term} = d,$ *then*

$$(\mathbf{rely}\, r) \Cap c \Cap d \sqsubseteq ((\mathbf{rely}\, r \cup r_1) \Cap (\mathbf{guar}\, r_2) \Cap c) \parallel ((\mathbf{rely}\, r \cup r_2) \Cap (\mathbf{guar}\, r_1) \Cap d).$$

Proof. The proof of this law uses Lemma 2 (rely-parallel-guar) to introduce parallel, in combination with applications of the interchange property between weak conjunction and parallel (54). First we show,

$$(\mathbf{rely}\, r) \Cap c \sqsubseteq ((\mathbf{rely}\, r) \Cap c) \parallel ((\mathbf{guar}\, r) \Cap \mathbf{term}) \tag{60}$$

as follows.

$$\begin{aligned}
(\mathbf{rely}\, r) \Cap c &= ((\mathbf{rely}\, r) \parallel (\mathbf{guar}\, r)) \Cap (c \parallel \mathbf{term}) &&\text{by (50) and assumption} \\
&\sqsubseteq ((\mathbf{rely}\, r) \Cap c) \parallel ((\mathbf{guar}\, r) \Cap \mathbf{term}) &&\text{by (54)}
\end{aligned}$$

Property (60) is used twice in the following proof of the law.

$$\begin{aligned}
&(\mathbf{rely}\, r) \Cap c \Cap d \\
\sqsubseteq\ &(\mathbf{rely}\, r \cup r_1) \Cap c \Cap (\mathbf{rely}\, r \cup r_2) \Cap d &&\text{by (52) and (49)} \\
\sqsubseteq\ &(((\mathbf{rely}\, r \cup r_1) \Cap c) \parallel ((\mathbf{guar}\, r \cup r_1) \Cap \mathbf{term})) \Cap \\
&(((\mathbf{guar}\, r \cup r_2) \Cap \mathbf{term}) \parallel ((\mathbf{rely}\, r \cup r_2) \Cap d)) &&\text{applying (60) twice} \\
\sqsubseteq\ &((\mathbf{rely}\, r \cup r_1) \Cap c \Cap (\mathbf{guar}\, r \cup r_2) \Cap \mathbf{term}) \parallel \\
&((\mathbf{guar}\, r \cup r_1) \Cap \mathbf{term} \Cap (\mathbf{rely}\, r \cup r_2) \Cap d) &&\text{by (54)} \\
\sqsubseteq\ &((\mathbf{rely}\, r \cup r_1) \Cap (\mathbf{guar}\, r_2) \Cap c) \parallel \\
&((\mathbf{rely}\, r \cup r_2) \Cap (\mathbf{guar}\, r_1) \Cap d) &&\text{by (42) and assumption}
\end{aligned}$$

$\square$

Because $[p,\ q_1 \cap q_2] = [p,\ q_1] \Cap [p,\ q_2]$ and specification commands satisfy the proviso conditions in Law 3, this law can be applied to refine a specification to a parallel composition.

10   Stuttering and Mumbling

Above, both tests and atomic steps have been treated as subsets of commands, where both subsets form Boolean algebras. In relational algebra approaches, it is common to identify tests as the subset of relations, each of which is a subset of the identity relation, i.e. a test that the state is in a set of states p is represented by the relation $p \lhd \mathsf{id}$, in which id is the identity relation on states and $p \lhd r$ is the restriction of the relation r so that its domain is contained in p. One approach (considered but not taken) was to identify tests with a subset of atomic program steps, i.e. a test that the state is in a set of states p could be represented by the command $\pi(p \lhd \mathsf{id})$. One consequence of this is that the conditional command **if** $\varnothing$ **then** c **else nil** would be encoded as

$$\pi(\varnothing \lhd \mathsf{id})\,;\, c \sqcap \pi(\overline{\varnothing} \lhd \mathsf{id})\,;\, \mathbf{nil} = \top\,;\, c \sqcap \pi(\mathsf{id}) = \pi(\mathsf{id})$$

That is, this conditional is defined to take a stuttering step, $\pi(\mathsf{id})$. That would appear to rule out the optimisation of implementing this conditional by **nil**. The usual approach to handle this issue in program semantics is to consider programs equivalent if their sets of traces are equivalent modulo finite stuttering, i.e. if all finite contiguous sequences of stuttering steps are removed from every trace, the sets of traces of the programs are the same. While this is straightforward to specify in a trace semantics, it is problematic to express it as more abstract algebraic properties of programs.

The approach we have taken is to treat tests as *instantaneous* commands that either succeed or fail and hence the set of test commands is essentially disjoint from the set of atomic step commands—they do have one element in common, $\tau(\varnothing) = \pi(\varnothing) = \epsilon(\varnothing) = \mathsf{T}$. To handle the instantaneous nature of tests we need additional axioms that show how tests and atomic steps combine. For program steps these axioms are,

$$\tau(p)\,;\,\pi(r) = \pi(p \lhd r) \tag{61}$$

$$\pi(r \rhd p)\,;\,\tau(p) = \pi(r \rhd p) \tag{62}$$

where $r \rhd p$ is the restriction of the relation r so that its range is contained in the set of states p.

Our specification command $[p,\ q]$ implicitly allows for finite stuttering in the sense that if c and d are semantically equivalent modulo finite stuttering then c refines a specification $[p,\ q]$ if and only if d refines $[p,\ q]$. In our approach c and d are not necessarily equal, but they do both refine the same specification. This is similar to the situation in sequential refinement where, although Quick Sort and Merge Sort both refine a suitable specification of sorting, they are not equivalent programs (neither refines the other) because Merge Sort is a stable sort but Quick Sort is not.

A second form of specification is $\langle p, q \rangle$. It performs an update that satisfies q atomically, provided the state before the update satisfies the precondition p. It allows finite stuttering steps before and after the update. The finite stuttering is represented by the command **idle** (63) that performs only a finite number of program steps that do change the state, (i.e. each program step it executes satisfies the identity relation, id, between its before and after states). The command $opt(q)$ performs the update q in a single atomic program step, $\pi(q)$, but if the before state σ is such that $(\sigma, \sigma) \in q$, q is satisfied by doing no step at all and hence the definition of $opt(q)$ allows a test for this case as an alternative. That allows optimisations like replacing an atomic update that does not change the state by a test (e.g. $opt(\mathsf{id}) = \pi(\mathsf{id}) \sqcap \tau(\{\sigma \mid (\sigma, \sigma) \in \mathsf{id}\}) = \pi(\mathsf{id}) \sqcap \mathbf{nil} \sqsubseteq \mathbf{nil}$).

$$\mathbf{idle} \mathrel{\hat{=}} (\mathbf{guar}\ \mathsf{id}) \Cap \mathbf{term} \tag{63}$$

$$opt(q) \mathrel{\hat{=}} \pi(q) \sqcap \tau(\{\sigma \mid (\sigma, \sigma) \in q\}) \tag{64}$$

$$\langle p, q \rangle = \mathbf{idle}\,;\{p\}\,;\,opt(q)\,;\,\mathbf{idle} \tag{65}$$

Note that in (65) the precondition must hold in the state immediately before the update. The **idle** commands before and after the optional update explicitly allow finite stuttering.

As well as specifications implicitly allowing for finite stuttering, our encodings of programming language commands and expression evaluation explicitly allow for finite stuttering. This allows one to consider transformations between programming language commands that are finite stuttering equivalent. The main advantage of this approach is that we are able handle finite stuttering algebraically and do not have to resort to the semantic model to handle refinement proofs reliant on finite stuttering equivalence.

Another equivalence used in trace semantics is mumbling equivalence [4], whereby a program with a trace containing two consecutive program steps $(\pi(r_1) \; ; \; \pi(r_2))$ also admits a trace with these two steps replaced by a single program step $\pi(r_1 \, \mathbin{\raise1pt\hbox{$\scriptstyle 9$}} \, r_2)$, where "$\mathbin{\raise1pt\hbox{$\scriptstyle 9$}}$" is relational composition. Again, while this is straightforward to represent in a trace semantics, it is problematic to express it as algebraic properties of programs. Again, in our approach a specification $[p, \; q]$ that admits an implementation that uses the two steps also admits an implementation that uses just one (combined) step.

The major advantage of treating finite stuttering and mumbling in the way we do is that it we can reason algebraically about the properties of commands, rather than having to prove properties in the semantic model whenever properties dependent on finite stuttering or mumbling arise. Our approach is more abstract because all properties shown to hold for an algebra, hold for any semantic model of the algebra.

11   Fairness

Our parallel operator has no fairness assumption but because we can explicitly refer to environment steps, we can encode a form of fairness that corresponds to minimal progress (i.e. if a process is always able to do a program step, it will eventually perform that step) within our algebra as a command, **fair**, that never allows its environment to perform an infinite number of steps in a row [15].

$$\mathbf{fair} \mathrel{\widehat{=}} (\epsilon^\star \; ; \; \pi)^\omega \; ; \; \epsilon^\star \tag{66}$$

Weak conjoining **fair** with a command c, i.e. $c \mathbin{\text{\small ⋒}} \mathbf{fair}$, denotes the fair execution of c and hence we may reason algebraically about the fair execution of c in isolation without concerning ourselves with a parallel process or a fair parallel operator. This gives a separation of concerns of fairness and parallel. For example, fair execution of **term** (39) only executes a finite number of steps.

$$
\begin{aligned}
\mathbf{term} \mathbin{\text{\small ⋒}} \mathbf{fair} &= (\epsilon^\omega \; ; \; \pi)^\star \; ; \; \epsilon^\omega \mathbin{\text{\small ⋒}} (\epsilon^\star \; ; \; \pi)^\omega \; ; \; \epsilon^\star \\
&= (\epsilon^\star \; ; \; \pi)^\star \; ; \; \epsilon^\star \\
&= (\epsilon \sqcap \pi)^\star && \text{by (11)} \\
&= \alpha^\star
\end{aligned}
$$

Hence if **term** $\sqsubseteq c$ then $\alpha^* = $ **term** $\pitchfork$ **fair** $\sqsubseteq c \pitchfork$ **fair**, i.e. fair execution of c only performs a finite number of steps.

A fair parallel operator can be defined using our existing parallel operator by requiring each operand of parallel to be fairly executed. The properties of fair parallel can then be derived from the existing parallel operator algebraically (see [15] for details).

12   Localisation

Like other programming constructs, local variable scopes can be defined in terms of more primitive operators and commands. The primitive operator at the core of local variable reasoning is the localisation operator $(\exists_x)$ that takes a variable x from a finite set of variables V and a command c, and effectively removes all restrictions on the variable x (but not on the other variables). It can be thought of as generalisation of the existential quantification operator of predicate calculus to an operator applicable to commands.

Given a state space $\Sigma \mathrel{\widehat{=}} V \to S$, and a variable $x \in V$, we say that states σ and σ' are *equivalent except for variable* x, written $\sigma \approx_x \sigma'$, when they have identical values for all variables other than x, i.e, $(\forall y \in V - \{x\} \cdot \sigma(y) = \sigma'(y))$. Given a set of states p over the state space Σ, we then have that the set of states $(\exists x \cdot p)$ contains state σ if and only if there exists a $\sigma' \in p$ such that $\sigma \approx_x \sigma'$.

A similar quantification operator $(\exists_x)$ can be defined over commands, whose observable effect on the state space can be represented by sequences of states. The notation for *equivalence except for a variable* x extends point-wise to (possibly infinite) sequences of states, i.e. for sequences $s, s' \in \mathrm{seq}\Sigma$ we have $s \approx_x s'$ iff sequence s and s' are of the same length and for all indices i in the sequences $s_i \approx_x s_i'$. We then specify that $\exists_x c$ can produce sequence of states s if and only if there exists a sequence of states s' produced by c satisfying $s \approx_x s'$.

For tests, which are a lifting of sets of states to commands, the localisation operator is homomorphic to existential quantification on sets of states, i.e. $\exists_x \tau(p) = \tau(\exists_x p)$. Arbitrary commands generalise this.

First-order equational logic, with its notion of variables and quantifications, have been given an algebraic characterisation in Henkin, Monk and Tarski's cylindric algebras [18]. The similarities shared between the primitive localisation operator on commands $(\exists_x)$ and existential quantification in predicate logic, means that their theory of cylindric algebras can be generalised to apply to program reasoning. This allows for a greater reuse of existing mathematical theories than would be possible had we decided to axiomatise a local variable scope command directly. Axioms for localisations include elegant properties such as $\exists_x c \sqsubseteq c$, which describes the fact that localisation introduces nondeterminism, commutativity property $\exists_x \exists_y c = \exists_y \exists_x c$, and distributivity properties $\exists_x (c \sqcap d) = \exists_x c \sqcap \exists_x d$ (like distribution of existential over disjunction in logic) and $\exists_x (c \parallel (\exists_x d)) = \exists_x c \parallel \exists_x d$ (like distribution of existential over conjunction in logic), etc. It follows that $\exists_x c \sqsubseteq \exists_x d$, if $c \sqsubseteq d$. The localisation operator $(\exists_x)$ extends to finite sets of variables in the obvious way, e.g. $\exists_{\{x,y\}} c = \exists_x \exists_y c$.

The inclusion of a localisation operator enriches the algebra with the extra expressivity required to reason about frames. Given that id is the identity relation on states, for any finite set of variables Y we have that $\pi(\exists_Y \text{id})$ is the atomic program step that can only modify variables in Y. Conjoining a command c with the guarantee $\mathbf{guar}(\exists_Y \text{id})$ then limits the frame of the command c to be Y, and so we define

$$Y : c \mathrel{\widehat{=}} \mathbf{guar}(\exists_Y \text{id}) \mathbin{\text{\reflectbox{m}}} c$$

to be the command c constrained so that its program steps can only modify variables in finite variable-set Y. For variable x we write $\overline{x}$ for the complement of x in the entire set of variables V. In this way we have that $\overline{x} : c$ limits command c so that its program steps cannot change x.

The axiomatisation of the localisation operator is arguably simpler than that of a local variable scope, especially in the context of shared memory concurrency, where interference must be accounted for. The command $(\mathbf{local}\, x \bullet c)$ performs c within the local variable scope of x. When c is executed within the local variable scope of x, variable x is not subject to interference from the environment. This is enforced by conjoining c with the command $\mathbf{demand}\, \text{id}_x$, where id_x is the identity relation on just x (i.e. $\text{id}_x = \exists_{\overline{x}}\text{id}$) and for a relation r,

$$\mathbf{demand}\, r \mathrel{\widehat{=}} (\pi \sqcap \epsilon(r))^\omega.$$

Command $\mathbf{demand}\, \text{id}_x$ therefore constrains the environment steps of c to not change x, in a manner similar to the way a guarantee command constrains program steps. Outside of the local variable scope $(\mathbf{local}\, x \bullet c)$, another variable with the same name as x is also defined, however the non-local occurrence of x is not modified by the program steps taken by $(\mathbf{local}\, x \bullet c)$, and it is unconstrained by the environment steps taken by $(\mathbf{local}\, x \bullet c)$. The behaviour of $(\mathbf{local}\, x \bullet c)$ is thus defined to be

$$(\mathbf{local}\, x \bullet c) \mathrel{\widehat{=}} \overline{x} : (\exists_x (c \mathbin{\text{\reflectbox{m}}} \mathbf{demand}\, \text{id}_x)).$$

The localisation on the command with the scope, $c \mathbin{\text{\reflectbox{m}}} \mathbf{demand}\, \text{id}_x$, removes all constraints on variable x, and the frame $\overline{x}$ then limits the behaviour of the program steps so that they cannot change non-local x.

Properties of local variable scopes are derivable from the more elementary properties of localisations and the other primitive operators. For example, we have the following monotonicity property and variable introduction rule, in which "x, Y" as a frame abbreviates $\{x\} \cup Y$.

$$(\mathbf{local}\, x \bullet c) \sqsubseteq (\mathbf{local}\, x \bullet d) \qquad\qquad \text{if } c \sqsubseteq d$$
$$Y : c = (\mathbf{local}\, x \bullet x, Y : c) \qquad\qquad \text{if } c = \exists_x c \text{ and } x \notin Y$$

Having the capability to reuse variable names simplifies reasoning about recursive procedure calls. Also, this treatment of localisations avoids having to introduce and remove variables from the state space, which is awkward when defining interleavings of program and environment steps, which may have different local variable declarations.

13   Some Remaining Challenges

Progress Properties. Handling progress within the concurrent refinement algebra is one of the remaining challenges. We are able to encode interval temporal logic (ITL) [34] properties of sequences of states as commands using the following scheme adapted from [7], in which an ITL formula f is encoded as a command $\mathcal{I}(f)$, the execution of which can generate all possible sequences of states satisfying f. For an ITL formula consisting of a state predicate p, p must hold in the initial state and any behaviour is permitted after that (i.e. **chaos**). A formula f holds in the next state, $\bigcirc f$, if it holds after performing any single step. A formula f holds in some state eventually, $\Diamond f$, if it holds after some finite number of steps. A formula f holds for all states, $\Box f$, if it holds initially and then $\Box f$ holds in the next state, which we represent here as a least fixed point. Conjunction and disjunction of ITL formulae map to weak conjunction and non-deterministic choice, respectively.

ITL formula f	Encoding $\mathcal{I}(f)$
p	$\tau(p)$; **chaos**
$\bigcirc f$	α ; $\mathcal{I}(f)$
$\Diamond f$	$\alpha^{\star}$; $\mathcal{I}(f)$
$\Box f$	$\mu\, x \cdot (\mathcal{I}(f) \Cap (\alpha\, ;\, x \sqcap \mathbf{nil}))$
$f_1 \wedge f_1$	$\mathcal{I}(f_1) \Cap \mathcal{I}(f_2)$
$f_1 \vee f_1$	$\mathcal{I}(f_1) \sqcap \mathcal{I}(f_2)$

One use of temporal logic is to specify that a command c implements a specification s provided c is executed in a context that satisfies a temporal logic formula f. That can be expressed as

$$s \sqsubseteq c \Cap \mathcal{I}(f)$$

that is, if c is restricted to executions that satisfy the temporal logic formula f, it implements s. The use of weak conjunction here (rather than strong conjunction) ensures that any aborting behaviour of c is not masked by the non-aborting encoding of ITL formulae. Temporal logic restrictions are commonly needed to show termination, for example, a dequeue operation that waits for the queue to be non-empty only terminates if eventually the queue is non-empty. Research on incorporating temporal logic assumptions to show termination for rely/guarantee concurrency is in its early stages.

Exclusive Access to Resources. Hoare introduced a command of the form **with** x **do** c to represent execution of the command c with the process having exclusive access to the resource x [20]. For example, a resource may be a shared variable x^4 and the **with** command effectively locks x for the duration of c. This command can be viewed as a specification of the desirable atomicity behaviour

---

[4] More generally it can be a set of variables.

for x and may be implemented using locks or other mechanisms to ensure mutual exclusion. It has an interesting interaction with rely and guarantee conditions in that it strengthens the rely within c to include the fact that x is not changed by the environment—normally relies can only be weakened (49)—and it weakens any guarantee on x to only need to hold end-to-end over c, rather than for every program step of c—normally guarantees can only be strengthened (42); see [16] for more details. Our research on refining concurrent programs has focused on the challenge of refining specifications to non-blocking algorithms and hence has not needed the abstraction of exclusive access to a resource, but research on incorporating such an abstraction is important for a comprehensive theory.

Data Refinement. Data refinement allows operations to be specified on an abstract data structure, (e.g. a bounded queue), but implemented using a more concrete data structure, (e.g. a circular buffer with read and write indices), with the two representations being linked by a coupling invariant. Morgan [32] uses an approach to data refinement that first augments the abstract operations with the concrete data structure, then refines the operations to utilise the concrete structure rather than the abstract structure (making use of the coupling invariant to accomplish this), and finally when the abstract structure is essentially an auxiliary variable, it may be eliminated. In the context of concurrency, the same general approach can be used but it needs to be adapted to handle the more general nature of commands in the concurrent refinement algebra.

14   Conclusions

Our theory focuses on the algebraic properties of operations on a set of commands ($\mathcal{C}$) which includes subsets representing instantaneous tests ($\mathcal{T}$) and atomic steps ($\mathcal{A}$). It builds on standard mathematical and computing algebraic theories as follows:

complete distributive lattice: commands $\mathcal{C}$ form a lattice with meet $\sqcap$ and join $\sqcup$, least element **abort** and greatest element $\top$;

fixed points: least (μ) and greatest (ν) fixed points allow one to define iterations and recursion;

monoids: $\mathcal{C}$ with sequential composition and identity **nil** forms a monoid, and $\mathcal{C}$ with parallel composition and identity **skip** forms a commutative monoid;

semi-lattice: $\mathcal{C}$ with weak conjunction ($\Cap$) and identity **chaos** forms a bounded semi-lattice;

Boolean algebra: sets of states ($\mathbb{P}\,\Sigma$), binary relations ($\mathbb{P}(\Sigma \times \Sigma)$), tests ($\mathcal{T}$), atomic steps ($\mathcal{A}$), the set of all program step commands of the form $\pi(r)$, and the set of all environment step commands of the form $\epsilon(r)$, all form Boolean algebras;

synchronous algebra: the operators parallel composition ($\|$), weak conjunction ($\Cap$) and the lattice join ($\sqcup$) all form synchronous algebras over the set of commands $\mathcal{C}$ with atomic step commands $\mathcal{A}$;

cylindric algebras: cylindric algebras can be defined over sets of states, binary relations, tests, atomic steps and commands to give a (generalised) notion of existential quantification for each of these.

As one can see we have been able to make extensive reuse of these algebras and their rich arrays of properties and that affords a simpler path to mechanisation of proofs in the theory. The hierarchy of theories has been progressively mechanised in the Isabelle/HOL interactive theorem prover. The hierarchy has been organised so that lemmas are proven at the level of the hierarchy that contains (just) the necessary definitions and properties; that affords greater opportunity of theory reuse.

The algebras separate out the properties of the individual operators but of course we also need to define how the operators interact with one another. All the operators ("$;$", "$\|$", "$\Cap$", and "$\exists_x$") distribute over non-empty non-deterministic choices and hence each of these operators is monotonic in its arguments. Combinations of operators like sequential and parallel composition do not have general distributive laws, rather they have weak interchange laws (24). Weak conjunction affords weak interchange laws with both sequential (53) and parallel composition (54).

Our theory has been successful to the extent that we have been able to encode all of the standard rely/guarantee refinement laws of Jones, in the process generalising many of the laws in the manner in which they handle expressions, including conditions. Our extensions include an atomic specification command, $\langle p, q \rangle$, that allows one to specify operations on a data structure that must appear to be atomic, and explicit handling of local variables. The latter is seen as our path to handling data refinement.

Schellhorn et al. have developed a rely/guarantee theory RGITL encoded in interval temporal logic [36]. Because we can encode interval temporal logic within our theory, it is at least as expressive as RGITL. However, the encoding of rely/guarantee in RGITL strictly alternates between environment and program steps, and thus a single environment step in RGITL corresponds to zero or more environment steps in our approach, and hence a rely condition in RGITL is the reflexive, transitive closure of the rely condition used here. Any finite sequence of environment steps in our approach (ϵ^*) is represented by a single step within RGITL, and hence our approach can make more fine-grained distinctions.

Acknowledgements. This research was supported by Discovery Grant DP190102142 from the Australian Research Council (ARC). Thanks are due to Joakim von Wright for introducing us to program algebra and Robert Colvin, Brijesh Dongol, Cliff Jones, Patrick Meiring, Kim Solin, Georg Struth, Andrius Velykis, and Kirsten Winter, for their input on ideas presented here.

References

1. Aczel, P.H.G.: On an inference rule for parallel composition. Private communication to Cliff Jones (1983). http://homepages.cs.ncl.ac.uk/cliff.jones/publications/MSs/PHGA-traces.pdf

2. Back, R.-J.R.: Correctness preserving program refinements: proof theory and applications. Tract 131, Mathematisch Centrum, Amsterdam (1980)
3. Back, R.-J.R., von Wright, J.: Refinement Calculus: A Systematic Introduction. Springer, New York (1998)
4. Brookes, S.: A semantics for concurrent separation logic. Theoret. Comput. Sci. **375**(1–3), 227–270 (2007)
5. Cohen, E.: Separation and reduction. In: Backhouse, R., Oliveira, J.N. (eds.) MPC 2000. LNCS, vol. 1837, pp. 45–59. Springer, Heidelberg (2000). https://doi.org/10.1007/10722010_4
6. Coleman, J.W., Jones, C.B.: A structural proof of the soundness of rely/guarantee rules. J. Log. Comput. **17**(4), 807–841 (2007)
7. Colvin, R.J., Hayes, I.J., Meinicke, L.A.: Designing a semantic model for a wide-spectrum language with concurrency. Formal Aspects Comput. **29**, 853 875 (2016)
8. de Roever, W.-P.: Concurrency Verification: Introduction to Compositional and Noncompositional Methods. Cambridge University Press, Cambridge (2001)
9. Dijkstra, E.W.: Guarded commands, nondeterminacy, and a formal derivation of programs. CACM **18**, 453–458 (1975)
10. Dijkstra, E.W.: A Discipline of Programming. Prentice-Hall, Upper Saddle River (1976)
11. Dingel, J.: A refinement calculus for shared-variable parallel and distributed programming. Formal Aspects Comput. **14**(2), 123–197 (2002)
12. Dongol, B., Hayes, I.J., Meinicke, L.A., Struth, G.: Cylindric kleene lattices for program construction. In: Hutton, G. (ed.) Mathematics of Program Construction 2019. LNCS. Springer, Cham, October 2019 (2019)
13. Hayes, I.J.: Generalised rely-guarantee concurrency: an algebraic foundation. Formal Aspects Comput. **28**(6), 1057–1078 (2016)
14. Hayes, I.J., Colvin, R.J., Meinicke, L.A., Winter, K., Velykis, A.: An algebra of synchronous atomic steps. In: Fitzgerald, J., Heitmeyer, C., Gnesi, S., Philippou, A. (eds.) FM 2016. LNCS, vol. 9995, pp. 352–369. Springer, Cham (2016). https://doi.org/10.1007/978-3-319-48989-6_22
15. Hayes, I.J., Meinicke, L.A.: Encoding fairness in a synchronous concurrent program algebra. In: Havelund, K., Peleska, J., Roscoe, B., de Vink, E. (eds.) Formal Methods. Lecture Notes in Computer Science, pp. 222–239. Springer International Publishing, Cham (2018)
16. Hayes, I.J.: Some challenges of specifying concurrent program components. In: Derrick, J., Dongol, B., Reeves, S. (eds.), Proceedings 18th Refinement Workshop, Electronic Proceedings in Theoretical Computer Science, Oxford, UK, 18th July 2018, vol. 282, pp. 10–22. Open Publishing Association, October 2018
17. Hayes, I.J., Meinicke, L.A., Winter, K., Colvin, R.J.: A synchronous program algebra: a basis for reasoning about shared-memory and event-based concurrency. Formal Aspects Comput. **31**(2), 133–163 (2019)
18. Henkin, L., Monk, J.D., Tarski, A.: Cylindric Algebras, Part I. Studies in logic and the foundations of mathematics, vol. 64. North-Holland Pub. Co., New York (1971)
19. Hoare, C.A.R.: An axiomatic basis for computer programming. Commun. ACM **12**(10), 576–580, 583 (1969)
20. Hoare, C.A.R.: Towards a theory of parallel programming. In: Operating System Techniques, pp. 61–71. Academic Press (1972)
21. Hoare, C.A.R., et al.: Laws of programming. Commun. ACM **30**(8), 672–686 (1987). Corrigenda: CACM **30**(9), 770

22. Hoare, C.A.R., He, J.: Unifying Theories of Programming. Prentice Hall, London (1998)
23. Hoare, C.A.R., Möller, B., Struth, G., Wehrman, I.: Concurrent Kleene algebra and its foundations. J. Log. Algebr. Program. **80**(6), 266–296 (2011)
24. Jones, C.B.: Development methods for computer programs including a notion of interference. Ph.D. thesis, Oxford University, June 1981. Available as: Oxford University Computing Laboratory (now Computer Science) Technical Monograph PRG-25
25. Jones, C.B.: Specification and design of (parallel) programs. In: Proceedings of IFIP 1983, pp. 321–332, North-Holland (1983)
26. Jones, C.B.: Tentative steps toward a development method for interfering programs. ACM ToPLaS **5**(4), 596–619 (1983)
27. Kozen, D.: Kleene algebra with tests. ACM Trans. Prog. Lang. Sys. **19**(3), 427–443 (1997)
28. Meinicke, L.A., Hayes, I.J.: Handling localisation in rely/guarantee concurrency: an algebraic approach. arXiv:1907.04005 [cs.LO] (2019)
29. Milner, A.J.R.G.: Communication and Concurrency. Prentice-Hall, Upper Saddle River (1989)
30. Milner, R.: Calculi for synchrony and asynchrony. Theoret. Comput. Sci. **25**(3), 267–310 (1983)
31. Morgan, C.C.: The specification statement. ACM Trans. Prog. Lang. Sys. **10**(3), 403–419 (1988)
32. Morgan, C.C.: Programming from Specifications, 2nd edn. Prentice Hall, Upper Saddle River (1994)
33. Morris, J.M.: A theoretical basis for stepwise refinement and the programming calculus. Sci. Comput. Program. **9**(3), 287–306 (1987)
34. Moszkowski, B.C.: Executing Temporal Logic Programs. Cambridge University Press, Cambridge (1986)
35. Prisacariu, C.: Synchronous Kleene algebra. J. Log. Algebr. Program. **79**(7), 608–635 (2010)
36. Schellhorn, G., Tofan, B., Ernst, G., Pfähler, J., Reif, W.: RGITL: a temporal logic framework for compositional reasoning about interleaved programs. Ann. Math. Artif. Intell. **71**(1–3), 131–174 (2014)
37. Schwarz, J.: Generic commands–a tool for partial correctness formalisms. Comput. J. **20**(2), 151–155 (1977)
38. von Wright, J.: Towards a refinement algebra. Sci. Comput. Program. **51**, 23–45 (2004)

UTP Semantics of a Calculus for Mobile Ad Hoc Networks

Xi Wu[1], Huibiao Zhu[2(✉)], and Wanling Xie[3]

[1] The University of Sydney, Sydney, Australia
[2] East China Normal University, Shanghai, China
hbzhu@sei.ecnu.edu.cn
[3] Nanjing University of Aeronautics and Astronautics, Nanjing, China

Abstract. The mCWQ calculus was recently proposed for describing the features of local broadcast and mobility in Mobile Ad Hoc Networks (MANETs), focusing on the quality of wireless communications. In this paper, we investigate the denotational semantics for mCWQ calculus, whose behaviour is composed of the behaviours of subnetworks. A trace variable tr is introduced to record the communications among wireless nodes as well as the time points when the communications happen. A set of algebraic laws, especially the laws about the communications with quality binders, are also explored based on the formalized model.

1 Introduction

To cater for the need for rapid development of Cyber Physical Systems (CPS) [11], Mobile Ad Hoc Networks (MANETs) [2,4] have drawn a great deal of attention recently, from both industry and academia. The interesting features of MANETs are local broadcast, node mobility and time consumption. In the literature, many research efforts are devoted to define a formal calculus and rigorous methodological foundations for modeling and reasoning about MANETs, focusing on these features, for example [5,7,13,14]. All of these calculi assume that communications in MANETs are reliable. However, deployment constraints and node movement of MANETs lead to a highly dynamical topology with wireless links to be broken, which may result in abnormalities and thus decrease the communication quality of service provided by the system. Therefore, it is of significant importance to ensure that wireless nodes in MANETs can still behave in a reasonable manner (e.g., continue to operate in a meaningful way) even though they are in an unreliable communication network due to node movement.

mCWQ calculus (Integrating a Calculus with Mobility and Quality for Wireless Networks) [21] was recently proposed based on our previous work [20,22]. It combines wireless local broadcast together with a quality predicate to enforce a robustness consideration on MANETs, and provide default values to make node behaviours reasonable. The topological structure is considered at the network level, incorporating time and mobility functions to capture the dynamic changes in the topology. Furthermore, a labeled transition semantics is developed to

© Springer Nature Switzerland AG 2019
P. Ribeiro and A. Sampaio (Eds.): UTP 2019, LNCS 11885, pp. 198–216, 2019.
https://doi.org/10.1007/978-3-030-31038-7_10

enable transitions, in which nodes may communicate with each other, change their mobility patterns (e.g., mobility function and timeout) or delay for some time units. However, the denotational semantics and algebraic laws of mCWQ calculus are not taken into consideration in [21].

Hoare and He advocate three different styles of mathematical representations, including operational semantics [17], denotational semantics [23] and algebraic semantics [9] in their Unifying Theories of Programming (UTP) [8]. Denotational semantics provides mathematical meanings to programs, while algebraic semantics fits well with symbolic calculation of parameters and structures of an optimal design. As far as we know, operational semantics is the commonly adopted semantics in reasoning about networks [10,12], whereas there are fewer researches on applying denotational semantics and algebraic semantics in networks. Providing these two semantics for networks may give a better and precise understanding of MANETs from mathematical perspective, and guide us to investigate more interesting properties of the networks, which create the motivation of this paper.

In this paper, we propose the denotational semantics for mCWQ calculus and deduce some interesting properties of the MANET system. In our semantic model, we give an observation tuple and introduce a variable tr to record the communications among nodes. Based on the formalized denotational semantics, we also investigate a set of algebraic laws, especially focusing on the communication between sender and receiver with binders.

The remainder of this paper is organized as follows. A review of the mCWQ calculus, including its syntax and mobility model, is given in Sect. 2. We investigate the semantic model of mCWQ calculus and healthiness conditions that a program should satisfy in Sect. 3, while in this section we also explore the denotational semantics of mCWQ calculus using the UTP approach. Section 4 presents some interesting algebraic laws of MANETs, especially focusing on the parallel expansion laws between sender and receiver with binders. Section 5 concludes the paper and presents the future work.

2 Review of the mCWQ Calculus

mCWQ calculus [21] was recently proposed for modeling and reasoning about MANETs and its applications. It uses quality predicates and default values to ensure that sensor nodes can behave in a reasonable manner (e.g., by using approximate values to continue their work when the ideal behavior of a node fails) even though they are in an unreliable communication network caused by node movement. In this section, we briefly review the syntax and mobility model of the mCWQ calculus.

The calculus is presented via a two-level syntax, including process level and network level, presented in Table 1. We employ P to range over the set of processes, N the set of nodes, Val the set of values, Var the set of variables and C the set of channels. We use the set $\mathcal{I}_n$ to denote node identities (or names), where n_1, n_2, ... range over $\mathcal{I}_n$. In order to decide whether a node is inside the transmission range of another, we define a distance function D, which takes two locations as inputs and returns the distance between these two locations. The transmission radius of a node is defined as a partial function $Rad : \mathcal{I}_n \times C \hookrightarrow \mathbf{R}_0^+$, taking a node identity and a channel name as parameters to calculate the radius of this node when it uses some channel to broadcast messages.

Processes. A process can be an inert process nil or an action prefixing $Act.P$. The process $P_1 \| P_2$ means that two processes run in parallel inside one node. A process can perform three kinds of actions: $c!v$, σ and b. The action $c!v$ denotes broadcasting a value $v \in Val$ via channel $c \in C$. In reality, broadcasting also uses a channel as wireless radio frequency to send messages, thus we write the broadcast channel c explicitly. Besides, as one novelty of mCWQ calculus, a time delay $\sigma \in \mathbf{R}^+$ is incorporated in the process level so that movement patterns can be given through a general mobility model depending on the time elapse.

Table 1. The Syntax of mCWQ

Processes:
$$P ::= \mathsf{nil} \mid Act.P \mid P\|P \mid \mathsf{case}\ e\ \mathsf{of}\ \mathsf{some}(x) : P\ \mathsf{else}\ P$$
$$Act ::= c!v \mid \sigma \mid b \qquad\qquad b ::= c?x \mid \&_q(b, ..., b)$$
Networks: *Function*:
$$N ::= 0 \mid n[P]_T^f \mid N\|N \qquad F ::= D(\overrightarrow{l}, \overrightarrow{l})$$

Another important and interesting thing in mCWQ calculus is the binder b, which is used to specify that a process can continue if the quality predicate is satisfied. The binder is first proposed in the Quality Calculus [15]. In the simplest case, it is a corresponding reception of a broadcasting action, represented as $c?x$, which receives a value via channel c and binds it to the variable $x \in Var$. A complex binder is in the form of $\&_q(b_1, ..., b_n)$, where n is the total number of inputs and q is a quality predicate to be satisfied, indicating to continue the process when sufficient inputs have been received. The quality predicate $q \in \{\forall, \exists, m/n\}$ and the meanings of these three notations are as follows:

- $\forall$: all inputs are required, e.g., $\&_\forall(c_1?x_1, c_2?x_2, c_3?x_3)$ requires three sufficient inputs and it has the same effect as $\&_q(c_1?x_1, c_2?x_2, c_3?x_3)$ if $q(r_1, r_2, r_3)$ amounts to $r_1 \wedge r_2 \wedge r_3$.
- $\exists$: at least one input is required, e.g., $\&_\exists(c_1?x_1, c_2?x_2, c_3?x_3)$ requires one sufficient input to continue and it has the same effect as $\&_q(c_1?x_1, c_2?x_2, c_3?x_3)$ if $q(r_1, r_2, r_3)$ amounts to $r_1 \vee r_2 \vee r_3$.
- m/n: m sufficient inputs of all n inputs are required to be received, e.g., $\&_{2/3}(c_1?x_1, c_2?x_2, c_3?x_3)$ requires at least two sufficient inputs from channels.

Moreover, nested binders are also allowed, such as $\&_\forall(\&_\exists(c_1?x_1, c_2?x_2), c_3?x_3)$, which represents that input must be received both over the channel c_3 and over either the channel c_1 or c_2. However, it is possible that some variables in the binder do not get proper values, due to the corresponding inputs having not occurred. Therefore, we use expression e to represent *optional data*, which includes expressions some($\cdot$) and none. The expression some($\cdot$) represents the presence of some data and none the absence of data. Because we are not sure which variable has actually received values at some time, there comes a construct case e of some(x) : P_1 else P_2 in the process part. It is used to check whether an expression e can evaluate to some data or not. If it does, then we bind it to x and continue with P_1; otherwise, we continue with P_2.

Networks. Networks are collections of nodes running in parallel with the form of $N_1 \| N_2$, each of which is inductively defined by parallel composition of an empty network 0 and a wireless node written as $n[P]_T^f$. Each node is assigned a unique identity $n \in \mathcal{I}_n$, and runs some process P with a mobility function f and its timeout T. We use $\mathcal{E}$ to represent the movement trajectory models of node mobility, which allows nodes to move in a global area A within a global time $t \in \mathbf{R}^+$. Each node has an entity mobility function $f : \mathbf{R}^+ \rightarrow A$, thus the location at time t of each node is decided by $f(t)$. Each mobility function has a timeout $T \in \mathbf{R}^+ \cup \{\infty, \diamond\}$. Here, $T \in \mathbf{R}^+$ means that the mobility function will be updated by a new one at time $t = T$, $T = \infty$ represents that the current mobility function will never be changed, and $T = \diamond$ denotes that the mobility function will be changed at any time.

$$\mathcal{E}(\widetilde{l_i}, t_i) = \{(f, T) \mid f(t) = \widetilde{l_i} + \widetilde{v} \cdot (t - t_i), \text{ where } \widetilde{v} \in V, \ T \in \mathbf{R}^+, \ t \in [t_i, T]\}$$

A new mobility function and its timeout will be selected nondeterministically by the mobility model $\mathcal{E}$, which takes a pair $(\widetilde{l_i}, t_i)$ as input and returns a set of pairs (f, T) of follow-up mobility functions with their timeouts. $\widetilde{l_i}$ stands for the current location of a node (represented by a vector[1]), t_i for the current time. Here, $\widetilde{v}$ is the speed which is a constant in a pre-defined speed set V.

3 Denotational Semantics

In this section, we develop a denotational semantics for mCWQ calculus. This theory is an extension of reactive processes [6,19].

Alphabet with Observational Variables. We introduce two variables $\overleftarrow{st}$ and $\overrightarrow{st}$ into our semantics to indicate the initial state and the final state of the network during the current observation, where $st \in \{terminate, wait\}$. A network program has two execution states:

- *terminate* state: A process may complete all its executing actions and terminate successfully. $\overleftarrow{st} = terminate$ denotes that the predecessor of the process has terminated successfully and the current process can take the control, while $\overrightarrow{st} = terminate$ means the current process terminates successfully.

---

[1] We use the notation $\sim$ to represent a vector, and leave the notation $\rightarrow$ to denote a final variable in semantics model (e.g., a final state is denoted as $\overrightarrow{st}$).

- $\overleftarrow{wait}$ state: A process is waiting for taking the control from its environment. $\overleftarrow{st} = wait$ indicates the predecessor of the process is waiting, thus the current process cannot be scheduled, while $\overrightarrow{st} = wait$ represents that the current process is waiting, thus, the next process cannot be activated.

The observation of the mCWQ calculus can be represented by a tuple: $(\overleftarrow{time}, \overrightarrow{time}, \overleftarrow{st}, \overrightarrow{st}, \overleftarrow{tr}, \overrightarrow{tr}, \overleftarrow{\mathcal{F}}, \overrightarrow{\mathcal{F}}, \overleftarrow{\mathcal{T}}, \overrightarrow{\mathcal{T}})$ where,

- $\overleftarrow{time}$ and $\overrightarrow{time}$ are start point and end point of a time interval over which the observation is recorded. We use $\delta(time)$ to stand for the length of the time interval, which is considered as a non-negative integer: $\delta(time) =_{df} (\overrightarrow{time} - \overleftarrow{time})$
- $\overleftarrow{tr}$ denotes the initial trace and $\overrightarrow{tr}$ represents the final trace of a program. Thus, $\overrightarrow{tr} - \overleftarrow{tr}$ stands for the sequence of snapshots contributed by the program and the environment during the time interval.

The behavior of the network is described in terms of a trace of snapshots. A snapshot in the trace variable is expressed by a pair (t, evt) where t indicates the time when the event happens; evt denotes the event. We use the projection to select the component of the snapshots:

$$\pi_1((t, evt)) =_{df} t \qquad \pi_2((t, evt)) =_{df} evt$$

Besides, $\mathcal{F}$ represents the mobility function and $\mathcal{T}$ denotes the corresponding timeout for some node n. The mobility function $\mathcal{F}$ takes the network node identifier as the input and returns the current mobility function the node n uses, such as f_1 or f_n. For simplicity, we omit its parameter in our semantic models. Specifically, we use $\overleftarrow{\mathcal{F}}$ to represent the initial mobility function with the corresponding timeout $\overleftarrow{\mathcal{T}}$ and $\overrightarrow{\mathcal{F}}$ denotes the final mobility function with its corresponding timeout $\overrightarrow{\mathcal{T}}$.

Healthiness Conditions. We use N to identify a network program in mCWQ, which must satisfy the following healthiness conditions. (H1) declares that the variable tr cannot be shortened because it is used to record the execution trace of a program, and the end point of a time interval cannot be less than the start point.

$$(\text{H1}) \quad N = N \wedge Inv(tr, time)$$

where $Inv(tr, time) =_{df} \overleftarrow{tr} \preceq \overrightarrow{tr} \wedge \overleftarrow{time} \leq \overrightarrow{time}$. We use the notation $\preceq$ to illustrate that $\overleftarrow{tr}$ is a prefix of $\overrightarrow{tr}$.

Besides, as we mentioned earlier, a network program may in the state of waiting, such as a network node is waiting for receiving messages from others or a node is executing a delay action. If a network program N is asked to start in a waiting state of its pre-program, then the state, trace, mobility function and the corresponding timeout of N remain unchanged; i.e., it satisfies the healthiness condition (H2).

$$(\text{H2}) \quad N = \Pi \lhd (\overleftarrow{st} = wait) \rhd N$$

where $\Pi =_{df} (\overrightarrow{st} = \overleftarrow{st}) \wedge (\overrightarrow{tr} = \overleftarrow{tr}) \wedge (\overrightarrow{\mathcal{F}} = \overleftarrow{\mathcal{F}}) \wedge (\overrightarrow{\mathcal{T}} = \overleftarrow{\mathcal{T}}) \wedge (\overrightarrow{time} = \overleftarrow{time})$ and $N_1 \lhd b \rhd N_2 =_{df} b \wedge N_1 \vee \neg b \wedge N_2$. The definition of $\mathcal{H}$-function can be given as below:

$$\mathcal{H}(X) =_{df} \Pi \lhd \overleftarrow{st} = wait \rhd (X \wedge Inv(tr, time))$$

From the definition of $\mathcal{H}$-function, we can see that $\mathcal{H}(X)$ satisfies both the healthiness conditions (H1) and (H2). This function is able to be used in defining the denotational semantics for the mCWQ calculus.

3.1 Basic Commands

Firstly, we give the denotational semantics for the basic commands, including broadcast, receiving, delay and receiving with binder, in mCWQ calculus as follows. We use the notation **beh** to represent the network behaviour. In Definition 1, we give the network behaviour of a node who is executing a broadcast output action. Note that, the action prefixing inside a node can be converted into the sequential operation between network behaviours of actions and processes. Thus, we use the notation $\mathring{,}$ to represent the sequential operation between network behaviours, which has the same meaning as the one used between processes in traditional calculi [8].

Definition 1 (Broadcast).

$$beh(n[c!v.P]_T^f) =_{df} beh(Bro(n, f, T, c, v)) \mathring{,} beh(n[P]_T^{\mathcal{F}})$$
$$beh(Bro(n, f, T, c, v)) =_{df}$$

$$\mathcal{H} \begin{pmatrix} \overrightarrow{st} = terminate \ \wedge \ \delta(time) = 0 \ \wedge \ \overrightarrow{\mathcal{F}} = \overleftarrow{\mathcal{F}} \ \wedge \ \overrightarrow{\mathcal{T}} = \overleftarrow{\mathcal{T}} \ \wedge \\ \overleftarrow{\mathcal{F}} = f \ \wedge \ \overleftarrow{\mathcal{T}} = T \ \wedge \ \overrightarrow{tr} = \overleftarrow{tr} \,\widehat{}\, \langle (\overrightarrow{time}, n[c, f, T].v) \rangle \end{pmatrix}$$

We use **beh**$(Bro(n, f, T, c, v))$ to stand for the network behaviour of the broadcasting action. Broadcasting action does not consume time, thus the interval of the observation time equals to 0. The execution state of the whole program is $terminate$ and this action will not affect the mobility function and the corresponding timeout of the node. After the execution, the trace of the node will be added with a new snapshot $\langle (\overrightarrow{time}, n[c, f, T].v) \rangle$. Here, $\overrightarrow{time}$ stands for the time of the action happening, while $n[c, f, T].v$ represents the corresponding event, that is the node n broadcasts a message v via channel c using mobility function f. We use notation $t_1 \,\widehat{}\, t_2$ to denote the concatenation of traces t_1 and t_2.

In Definition 2, we discuss about the corresponding receive action. We suppose the mobility function that node n currently uses is f_1 and its corresponding timeout is T_1. The process $c?x.P$ represents that if the node receives a message from a sender via the channel c, the whole process will behave as P; otherwise, it will remain in a waiting state. We assume that there exists a message value m, which has the type that can be accepted via channel c. The notation **beh**$(BRecv(n, f_1, T_1, c, m))$ is used to represent the network behaviour of receiving. After a successful receive event, the value of free variable x in P will be replaced by the value of the message m, denoted as $P\{m/x\}$. Note that, waiting for receiving messages accompanies with time consuming. In mCWQ calculus, because we take node mobility into consideration, it is possible for a node to change its mobility function and timeout when it waits for receiving messages. Therefore, after receiving, its mobility function may not be the same as the initial one f_1.

Definition 2 (Receive).

$$beh(n[c?x.P]_{T_1}^{f_1}) =_{df} \exists m \in Val \wedge m \in Type(c) \bullet$$

$$\left(beh(\textbf{\textit{BRecv}}(n, f_1, T_1, c, m)) \, \fatsemi \, beh(n[P\{m/x\}]_{\mathcal{T}}^{\mathcal{F}})\right)$$

$$beh(\textbf{\textit{BRecv}}(n, f_1, T_1, c, m)) =_{df}$$

$$\mathcal{H} \left(\begin{array}{l} \overrightarrow{st} = terminate \ \wedge \ \delta(time) \geq 0 \ \wedge \ \exists n', f', T' \bullet \\[4pt] \left(\begin{array}{l} \overrightarrow{\mathcal{F}} = \overleftarrow{\mathcal{F}} \wedge \overrightarrow{\mathcal{T}} = \overleftarrow{\mathcal{T}} \ \wedge \ \overleftarrow{\mathcal{F}} = f_1 \ \wedge \ \overleftarrow{\mathcal{T}} = T_1 \ \wedge \\[4pt] D(\overrightarrow{\mathcal{F}}(\overrightarrow{time}), f'(\overrightarrow{time})) \leq Rad(n', c) \ \wedge \\[4pt] \overrightarrow{tr} = \overleftarrow{tr} ^\frown \langle (\overrightarrow{time}, n'[c, f', T'].m) \rangle \end{array} \right) \\[4pt] \lhd (\overrightarrow{time} \leq \overleftarrow{\mathcal{T}}) \rhd \\[4pt] \left(\begin{array}{l} \exists f_1, \ldots, f_n, T_1, \ldots, T_n, \forall i \in \{1, \ldots, n-1\} \bullet \\[4pt] T_i \in [\overleftarrow{time}, \overrightarrow{time}] \ \wedge \ \overleftarrow{\mathcal{F}} = f_1 \ \wedge \ \overleftarrow{\mathcal{T}} = T_1 \ \wedge \ \overrightarrow{\mathcal{F}} = f_n \wedge \overrightarrow{\mathcal{T}} = T_n \ \wedge \\[4pt] D(\overrightarrow{\mathcal{F}}(\overrightarrow{time}), f'(\overrightarrow{time})) \leq Rad(n', c) \ \wedge \\[4pt] (f_{i+1}, T_{i+1}) \in \mathcal{E}(f(T_i), T_i) \ \wedge \ \overrightarrow{tr} = \overleftarrow{tr} ^\frown \langle (\overrightarrow{time}, n'[c, f', T'].m) \rangle \end{array} \right) \\[4pt] \vee \\[4pt] \overrightarrow{st} = wait \ \wedge \ \delta(time) \geq 0 \wedge \overrightarrow{tr} = \overleftarrow{tr} \ \wedge \\[4pt] \left(\overrightarrow{\mathcal{F}} = \overleftarrow{\mathcal{F}} \wedge \overrightarrow{\mathcal{T}} = \overleftarrow{\mathcal{T}} \ \wedge \ \overleftarrow{\mathcal{F}} = f_1 \ \wedge \ \overleftarrow{\mathcal{T}} = T_1 \right) \\[4pt] \lhd (\overrightarrow{time} \leq \overleftarrow{\mathcal{T}}) \rhd \\[4pt] \left(\begin{array}{l} \exists f_1, \ldots, f_n, T_1, \ldots, T_n, \forall i \in \{1, \ldots, n-1\} \bullet \\[4pt] T_i \in [\overleftarrow{time}, \overrightarrow{time}] \ \wedge \ \overleftarrow{\mathcal{F}} = f_1 \ \wedge \ \overleftarrow{\mathcal{T}} = T_1 \ \wedge \\[4pt] \overrightarrow{\mathcal{F}} = f_n \wedge \overrightarrow{\mathcal{T}} = T_n \wedge (f_{i+1}, T_{i+1}) \in \mathcal{E}(f(T_i), T_i) \end{array} \right) \end{array} \right)$$

The network behaviours can be considered into two branches: (1) the node succeeds in receiving messages from a sender within the observation interval; (2) the node is always in the waiting state within the observation interval. For the first branch, the final execution state is *terminate* and the length of the time interval should not be negative. We suppose that the sender has an identifier n', mobility function f' and timeout T'. According to whether the end point of the observation interval is larger than the timeout point of the current mobility function of the node, the first branch can also be divided into two cases. If the end point of the observation interval (or the execution time point of the receiving action) is smaller than the timeout point of the initial mobility function, which means that the node does not change its mobility function before the receiving action happens, then the mobility function and the timeout keep unchanged. A snapshot $\langle(\overrightarrow{time}, n'[c, f', T'].m)\rangle$, including the instant $\overrightarrow{time}$ at which the event $n'[c, f', T'].m$ occurs and the event, is added in the end of the original trace. Note that, if the node can receive a message successfully, it means that this node is inside the transmission area of the sender, which can be restricted by the condition $D(\overrightarrow{\mathcal{F}}(\overrightarrow{time}), f'(\overrightarrow{time})) \leq Rad(n', c)$.

Definition 3 (Delay).

$$beh(n[\sigma.P]_{T_1}^{f_1}) =_{df} beh(\boldsymbol{Delay}(n, f_1, T_1, \sigma)) \,\fatsemi\, beh(n[P]_{T}^{\mathcal{F}})$$

$$beh(\boldsymbol{Delay}(n, f_1, T_1, \sigma)) =_{df}$$

$$\mathcal{H}\left|\begin{pmatrix} \overrightarrow{st} = terminate \,\wedge\, \delta(time) = \sigma \,\wedge \\ \left(\begin{pmatrix} \overrightarrow{time} < \overleftarrow{T} \,\wedge\, \overrightarrow{tr} = \overleftarrow{tr} \,\wedge\, \overrightarrow{\mathcal{F}} = \overleftarrow{\mathcal{F}} \,\wedge\, \overrightarrow{T} = \overleftarrow{T} \,\wedge \\ \overleftarrow{\mathcal{F}} = f_1 \,\wedge\, \overleftarrow{T} = T_1 \end{pmatrix} \\ \vee \\ \begin{pmatrix} \overrightarrow{time} \geq \overleftarrow{T} \,\wedge\, \overrightarrow{tr} = \overleftarrow{tr} \,\wedge \\ \exists f_1, \ldots, f_n, T_1, \ldots, T_n, \forall i \in \{1, \ldots, n-1\} \bullet \\ T_i \in [\overleftarrow{time}, \overrightarrow{time}] \,\wedge\, \overleftarrow{\mathcal{F}} = f_1 \,\wedge\, \overleftarrow{T} = T_1 \,\wedge \\ \overrightarrow{\mathcal{F}} = f_n \wedge \overrightarrow{T} = T_n \wedge (f_{i+1}, T_{i+1}) \in \mathcal{E}(f(T_i), T_i) \end{pmatrix} \right) \\ \vee \\ \overrightarrow{st} = wait \,\wedge\, \delta(time) < \sigma \,\wedge \\ \left(\begin{pmatrix} \overrightarrow{time} < \overleftarrow{T} \,\wedge\, \overrightarrow{tr} = \overleftarrow{tr} \,\wedge \\ \overrightarrow{\mathcal{F}} = \overleftarrow{\mathcal{F}} \,\wedge\, \overrightarrow{T} = \overleftarrow{T} \,\wedge\, \overleftarrow{\mathcal{F}} = f_1 \,\wedge\, \overleftarrow{T} = T_1 \end{pmatrix} \\ \vee \\ \begin{pmatrix} \overrightarrow{time} \geq \overleftarrow{T} \,\wedge\, \overrightarrow{tr} = \overleftarrow{tr} \,\wedge \\ \exists f_1, \ldots, f_n, T_1, \ldots, T_n, \forall i \in \{1, \ldots, n-1\} \bullet \\ T_i \in [\overleftarrow{time}, \overrightarrow{time}] \,\wedge\, \overleftarrow{\mathcal{F}} = f_1 \,\wedge\, \overleftarrow{T} = T_1 \,\wedge \\ \overrightarrow{\mathcal{F}} = f_n \wedge \overrightarrow{T} = T_n \wedge (f_{i+1}, T_{i+1}) \in \mathcal{E}(f(T_i), T_i) \end{pmatrix} \right) \end{pmatrix}\right.$$

However, if the end point of the observation interval (or the execution time point of the receiving action) is larger than the timeout point of the initial mobility function of the node, it means that the mobility function of the node has already changed before the receiving action happens. We assume that the node changes its mobility function $n - 1$ times from the starting point of the observation until the receiving action happens. Then, there must exist two sequences $f_1, ..., f_n$ and $T_1, ..., T_n$ representing the mobility function sequence and the timeout sequence, respectively. For any $i \in \{1, ..., n-1\}$, the elements in the sequences satisfy the conditions of $T_i \in [\overleftarrow{time}, \overrightarrow{time}]$ and $(f_{i+1}, T_{i+1}) \in \mathcal{E}(f(T_i), T_i)$. We can see that the initial mobility function the node uses is f_1 and its timeout is T_1; while the final mobility function the node uses at the end point of the observation interval is f_n and the corresponding timeout is T_n. Also, a snapshot $\overrightarrow{tr} = \overleftarrow{tr}^\frown \langle (\overrightarrow{time}, n'[c, f', T'].m) \rangle$ is added in the end of the trace, and the distance restriction $D(\overrightarrow{\mathcal{F}}(\overrightarrow{time}), f'(\overrightarrow{time})) \leq Rad(n', c)$ is satisfied.

For the second branch, the final execution state is wait and the length of the observation interval is still not negative. The trace of the network behaviour remains unchanged. We omit the explanations of this branch due to it being similar to the first one.

In Definition 3, we give the network behaviour of the delay action. We still suppose that the initial mobility function of node n is f_1 and the corresponding timeout is T_1. The executed process inside the node is $\sigma.P$, representing that after σ time units, the whole process inside the node behaves as P. Here, we use the notation $\mathbf{beh}(\mathbf{Delay}(n, f_1, T_1, \sigma))$ to stand for the network behaviour of a node that executes a delay action.

The network behaviour of delay action can also be considered into two cases: (1) the observation time interval equals to the delay time units, which means that the node finishes executing the delay action within the observation time interval; (2) the observation time interval is less than the required delay time units, that is the node still needs to delay more time units except the time interval of the observation.

In the first case, the final state of the network behaviour of the node is *terminate* and the node has already delayed for σ time units. During these σ time units, if the node doesn't change its mobility function (i.e., $\overrightarrow{time} < \overleftarrow{T}$), then the trace of this node keeps unchanged as well as its mobility function and the corresponding timeout. Otherwise, this node may change its mobility function greater than or equal to one time and there may exist two sequences $f_1, \ldots, f_n$ and $T_1, \ldots, T_n$ for mobility function sequence and corresponding timeout sequence, respectively. Whereas, the trace of the node remains unchanged.

For the second case, due to the time interval of the observation is less than the required delay time units, thus the final state of the program is wait, which means that it still needs to wait for more time units. In this case, the trace of the node is still unchanged. Similar with the first case, whether the mobility function of the node will change or not depends on the comparison of the end point of the observation and the corresponding timeout of the initial mobility function the node uses.

Definition 4 (Receive with Binder).

$$\boldsymbol{beh}(n[\&_q(b_1, ..., b_n).P]_T^f) =_{df} \boldsymbol{beh}(\boldsymbol{QRecv}(n, q, f, T, b_1, ..., b_n)) \, \boldsymbol{\mathring{,}} \, \boldsymbol{beh}(n[P]_T^{\mathcal{F}})$$

$$\boldsymbol{beh}(\boldsymbol{QRecv}(n, q, f, T, b_1, ..., b_n)) =_{df}$$

$$\left(\begin{array}{l} \exists \, \overleftarrow{st_1}, \ldots, \overleftarrow{st_n}, \overrightarrow{st_1}, \ldots, \overrightarrow{st_n}, \overleftarrow{tr_1}, \ldots, \overleftarrow{tr_n}, \overrightarrow{tr_1}, \ldots, \overrightarrow{tr_n}, \forall \, i \in \{1, \ldots, n\} \bullet \\ \overleftarrow{tr_i} = \overleftarrow{tr} \, \wedge \, \overleftarrow{st_i} = \overleftarrow{st} \, \wedge \, \boldsymbol{beh}_i[\overleftarrow{st_i}, \overrightarrow{st_i}, \overleftarrow{tr_i}, \overrightarrow{tr_i}/\overleftarrow{st}, \overrightarrow{st}, \overleftarrow{tr}, \overrightarrow{tr}] \, \wedge \, QMerge(q) \end{array} \right)$$

Definition 5 (Predicate for Binder).

$QMerge(q) \stackrel{df}{=}$

$$
\left(
\begin{array}{l}
\left(
\begin{array}{l}
q = \forall \;\wedge \\
\overrightarrow{st_1} = terminate \;\wedge\; \ldots \;\wedge\; \overrightarrow{st_n} = terminate \;\Rightarrow\; \overrightarrow{st} = terminate \;\wedge \\
\left(
\begin{array}{l}
\exists\, i \in \{1, \ldots, n\} \;\bullet\; \overrightarrow{st_i} = wait \;\wedge \\
\bigwedge_{j\in(\{1,\ldots,n\}-\{i\})} \overrightarrow{st_j} \in \{terminate, wait\} \;\Rightarrow\; \overrightarrow{st} = wait
\end{array}
\right) \;\wedge \\
\exists\, \omega \in (\overrightarrow{tr_1} - \overleftarrow{tr}) \;\oplus\; (\overrightarrow{tr_2} - \overleftarrow{tr}) \;\oplus\; \ldots \;\oplus\; (\overrightarrow{tr_n} - \overleftarrow{tr}) \;\bullet\; \overrightarrow{tr} = \overleftarrow{tr}{}^\frown \omega
\end{array}
\right) \\
\vee \\
\left(
\begin{array}{l}
q = \exists \;\wedge \\
\left(
\begin{array}{l}
\exists\, i \in \{1, \ldots, n\} \;\bullet\; \overrightarrow{st_i} = terminate \;\wedge \\
\bigwedge_{j\in(\{1,\ldots,n\}-\{i\})} \overrightarrow{st_j} \in \{terminate, wait\} \;\Rightarrow\; \overrightarrow{st} = terminate
\end{array}
\right) \;\wedge \\
\overrightarrow{st_1} = wait \;\wedge\; \ldots \;\wedge\; \overrightarrow{st_n} = wait \;\Rightarrow\; \overrightarrow{st} = wait \;\wedge \\
\exists\, \omega \in (\overrightarrow{tr_1} - \overleftarrow{tr}) \;\oplus\; (\overrightarrow{tr_2} - \overleftarrow{tr}) \;\oplus\; \ldots \;\oplus\; (\overrightarrow{tr_n} - \overleftarrow{tr}) \;\bullet\; \overrightarrow{tr} = \overleftarrow{tr}{}^\frown \omega
\end{array}
\right) \\
\vee \\
\left(
\begin{array}{l}
q = m/n \;\wedge \\
\left(
\begin{array}{l}
\exists\, j_1, \ldots, j_m \in \{1, \ldots, n\} \;\wedge\; j_1 \neq \ldots \neq j_m \;\bullet \\
\overrightarrow{st_{j_1}} = terminate \;\wedge \ldots \wedge\; \overrightarrow{st_{j_m}} = terminate \;\wedge \\
\bigwedge_{i\in(\{1,\ldots,n\}-\{j_1,\ldots,j_m\})} \overrightarrow{st_i} \in \{wait, terminate\} \;\Rightarrow\; \overrightarrow{st} = terminate
\end{array}
\right) \\
\wedge\;
\left(
\begin{array}{l}
\exists\, j_1, \ldots, j_k \in \{1, \ldots, n\} \;\wedge\; j_1 \neq \ldots \neq j_k \;\wedge\; 0 \le k < m \;\bullet \\
\overrightarrow{st_{j_1}} \in \{wait, terminate\} \;\wedge \ldots \wedge\; \overrightarrow{st_{j_k}} \in \{wait, terminate\} \;\wedge \\
\bigwedge_{i\in(\{1,\ldots,n\}-\{j_1,\ldots,j_k\})} \overrightarrow{st_i} = wait \;\Rightarrow\; \overrightarrow{st} = wait
\end{array}
\right) \\
\wedge\; \exists\, \omega \in (\overrightarrow{tr_1} - \overleftarrow{tr}) \;\oplus\; (\overrightarrow{tr_2} - \overleftarrow{tr}) \;\oplus\; \ldots \;\oplus\; (\overrightarrow{tr_n} - \overleftarrow{tr}) \;\bullet\; \overrightarrow{tr} = \overleftarrow{tr}{}^\frown \omega
\end{array}
\right)
\end{array}
\right)
$$

We will define the network behaviour of a node who executes receiving with a binder b in Definitions 4 and 5. According to the syntax of mCWQ calculus in Table 1, the binder b is used to describe that a node can execute the continuous process after some sufficient receiving actions are executed (or we can say that the quality predicate q in the binder b is satisfied). The binder b is composed of a simple receiving action and a complex receiving action with quality predicate, whose definition is given as follows:

$$b ::= c?x \mid \&_q(b_1, \ldots, b_n)$$

We have already given the network behaviour of the simple receiving action $c?x$ in Definition 2. For the complex binder with the form of $\&_q(b_1, \ldots, b_n)$, n is the total number of required inputs and q stands for a quality predicate to be satisfied, indicating to continue the process when sufficient inputs have been received. Here, we use the notation **beh(QRecv($n, q, f, T, b_1, \ldots, b_n$))** to represent the network behaviour of a node who executes receiving with the complex binders. The binder b in $\&_q(b_1, \ldots, b_n)$ can either be a simple receiving action $c?x$ or a complex receiving action $\&_q(b_1, \ldots, b_n)$ as well, and so we will deal with its network behaviour in two cases.

For the first case, we suppose all the binders b in $\&_q(b_1, ..., b_n)$ are of the form of the simple receiving actions. In this case, we can define the network behaviour of $\&_q(b_1, ..., b_n)$ depending on the network behaviour of receiving action given in Definition 2 and we will write $\mathbf{beh}(\mathbf{BRecv}(n, f_1, T_1, c, m))$ as $\mathbf{beh}_1, \ldots, \mathbf{beh}_n$ for simplicity. The quality predicate in the complex binder has three possible values, that is $q \in \{\forall, \exists, m/n\}$ and the corresponding meanings of these three notations are all inputs are required to be executed, at least one input is required to be executed and m sufficient inputs of all n inputs are required to be executed, respectively. We give the definition for the first case in Definition 4, in which the initial trace and final trace of all receiving actions are the same, and the corresponding network behaviours for these actions are denoted as $\mathbf{beh}_1, \ldots, \mathbf{beh}_n$. The predicate QMerge($q$) is used to merge the different traces produced by each binder according to different value of the quality predicate q. The definition of this predicate can be found in Definition 5.

According to the different values of q, the predicate QMerge(q) has three cases to be discussed in the denotational semantics. The descriptions of these three cases can be found below.

- When the quality predicate q equals $\forall$: All the inputs are required to be executed before the whole process to be continue. Thus, the final state of the complex binder will be *terminate* only if the final states (i.e., $\overrightarrow{st_1}, \ldots, \overrightarrow{st_n}$) of the network behaviours of all the required input actions (i.e., $\mathbf{beh}_1, \ldots, \mathbf{beh}_n$) are *terminate*. If there is at least one input action whose final state of the network behaviour is *wait*, then the final state of this complex binder should be *wait*.

- When the quality predicate q equals $\exists$: At least one input is required to be executed before the whole process can continue. Thus, the final state of the complex binder will be *terminate* if there is at least one required input, whose final state of the network behaviour (e.g., $\mathbf{beh}_i$) is *terminate*. If the final states of the network behaviours of all these required input are *wait*, then the final state of this complex binder should be *wait*.

- When the quality predicate q equals m/n: There are n inputs in total and at least m inputs are required before the whole process to be continue. Thus, the final state of the complex binder will be *terminate* if there are at least m required inputs, whose final state of the network behaviours (e.g., $\mathbf{beh}_{j_1}, \ldots, \mathbf{beh}_{j_m}$, where $j_1 \neq, \ldots, \neq j_m$) is *terminate*. If the amount of the inputs, whose final state of the network behaviour is *terminate*, is less than m of the total number n, then the final state of this complex receiving action should be *wait*.

We use the operator $\oplus$ to merge the traces produced by the network behaviours of n receiving actions, denoted as $tr_1 \oplus tr_2 \oplus \ldots \oplus tr_n$. Merging rules for the operator $\oplus$ are given in Appendix A.

As mentioned above, we discuss the first case of the network behaviour of the receiving action with binders, that is the binders b in this action are all composed of simple receiving action. Another case of the network behaviour of this action takes nested receiving actions in b into consideration. On one hand, we have already defined the network behaviour of the first case of the receiving action with binders using $\mathbf{beh}(\mathbf{QRecv}(n, q, f, T, b_1, ..., b_n))$ in Definition 4; on the other hand, receiving actions with nested binders can finally be expanded into simple receiving actions, thus, we use a recursive definition to define the network behaviour of the second case depending on Definition 4 and use $\mathbf{qbeh}_i$ to represent $\mathbf{beh}_i(\mathbf{QRecv}(n, q, f, T, b_1, ..., b_n))$ for simplicity. Then, the definition for the second case can be given by using $\mathbf{qbeh}_1, ..., \mathbf{qbeh}_n$ to replace $\mathbf{beh}_1, ..., \mathbf{beh}_n$ from Definition 4 and others remain unchanged.

3.2   Parallel Composition

In this section, we will give the semantics for the parallel composition of processes and networks. Here, P and Q represent two processes and $P\|Q$ stands for the parallel composition between two processes executed inside one network node. The network behaviour of this operation can be regarded as the parallel composition between the network behaviour of process P and the network behaviour of process Q, and we need to merge the traces produced by each of these two network behaviours. Details of the definition for process parallel composition can be found in Definition 6.

Definition 6 (Parallel Composition on Processes).

$$\mathbf{beh}(n[P\|Q]_T^f) =_{df} \mathbf{beh}(n[P]_T^f)\|\mathbf{beh}(n[Q]_T^f)$$

$$where,\ F\|G =_{df}$$

$$\left(\begin{array}{l} \exists \overleftarrow{st_1}, \overrightarrow{st_1}, \overleftarrow{st_2}, \overrightarrow{st_2}, \overleftarrow{tr_1}, \overrightarrow{tr_1}, \overleftarrow{tr_2}, \overrightarrow{tr_2} \bullet \overleftarrow{tr_1} = \overleftarrow{tr_2} = \overleftarrow{tr} \wedge \overleftarrow{st_1} = \overleftarrow{st_2} = \overleftarrow{st} \wedge \\ F[\overleftarrow{st_1}, \overrightarrow{st_1}, \overleftarrow{tr_1}, \overrightarrow{tr_1}/\overleftarrow{st}, \overrightarrow{st}, \overleftarrow{tr}, \overrightarrow{tr}] \wedge \\ G[\overleftarrow{st_2}, \overrightarrow{st_2}, \overleftarrow{tr_2}, \overrightarrow{tr_2}/\overleftarrow{st}, \overrightarrow{st}, \overleftarrow{tr}, \overrightarrow{tr}] \wedge Merge \end{array}\right)$$

This definition is given based on parallel-by-merge [3,8]. The predicates in the last two lines stand for the individual behaviours produced by F and G, respectively. The last predicate $Merge$ is used to merge the individual executed traces produced by two parallel branches according to the communication between them. The definition of the predicate $Merge$ is shown below, which is similar to the merge predicates for CSP and Circus in [1,16,18].

Definition 7 (Predicate for Parallel Composition on Processes).

$$Merge =_{df}$$

$$\left(\begin{array}{l} \overrightarrow{st_1} = terminate \ \wedge \ \overrightarrow{st_2} = terminate \ \Rightarrow \ \overrightarrow{st} = terminate \ \wedge \\ \overrightarrow{st_1} = wait \ \wedge \ \overrightarrow{st_2} \in \{terminate, wait\} \ \Rightarrow \ \overrightarrow{st} = wait \ \wedge \\ \overrightarrow{st_2} = wait \ \wedge \ \overrightarrow{st_1} \in \{terminate, wait\} \ \Rightarrow \ \overrightarrow{st} = wait \ \wedge \\ \exists u \in (\overrightarrow{tr_1} - \overleftarrow{tr}) \mid (\overrightarrow{tr_2} - \overleftarrow{tr}) \bullet \overrightarrow{tr} = \overleftarrow{tr}^\frown u \end{array}\right)$$

As shown in Definition 6, the final state of the parallel composition of F and G is decided by these two components. Only when the state of both these two components are *terminate*, the final state of the parallel composition is *terminate*; otherwise, if the state of any one component is *wait*, then the state of the parallel composition should be *wait*. The rules for merging the individual traces, which are used to describe the communication behaviours of the parallel composition, produced by components are given in Definition 8.

Definition 8 (Merging Rules for Traces).

$$tr_1 \mid tr_2 =_{df}$$

1. $\varepsilon \mid \varepsilon = \{\varepsilon\}$
2. $\varepsilon \mid \langle (t, evt) \rangle \widehat{\ } s = \{\langle (t, evt) \rangle \widehat{\ } s\}$
3. $\langle (t, evt) \rangle \widehat{\ } s \mid \varepsilon = \varepsilon \mid \langle (t, evt) \rangle \widehat{\ } s$
4. $\langle (t_1, evt_1) \rangle \widehat{\ } s \mid \langle (t_2, evt_2) \rangle \widehat{\ } u =$

$$
\begin{cases}
\langle (t_1, evt_1) \rangle \widehat{\ } (s \mid \langle (t_2, evt_2) \rangle \widehat{\ } u) & \text{if } t_1 < t_2, \\
\langle (t_2, evt_2) \rangle \widehat{\ } (\langle (t_1, evt_1) \rangle \widehat{\ } s \mid u) & \text{if } t_1 > t_2, \\
\langle (t_1, evt_1) \rangle \widehat{\ } (s \mid u) & \text{if } t_1 = t_2 \wedge evt_1 = evt_2, \\
\langle (t_1, evt_1) \rangle \widehat{\ } (s \mid \langle (t_2, evt_2) \rangle \widehat{\ } u) \cup \langle (t_2, evt_2) \rangle \widehat{\ } (\langle (t_1, evt_1) \rangle \widehat{\ } s \mid u) \\
\qquad\qquad\qquad\qquad\qquad\qquad\qquad\quad \text{if } t_1 = t_2 \wedge evt_1 \neq evt_2.
\end{cases}
$$

Merge result of two non-empty traces is according to their executed time. The first case describes that the executed time of the first snapshot of trace s is earlier than the executed time of the first snapshot of trace u, thus we should put the first snapshot of trace s into the trace of parallel composition; otherwise, if the executed time of the first snapshot of trace u is earlier than the executed time of the first snapshot of trace s, then we should put the first snapshot of trace u into the trace of parallel composition, which is shown in the second case. Besides, if the executed time of the first snapshot of trace s equals to the executed time of the first snapshot of trace u, then the order between these two snapshots should be decided according to whether the communication event is the same or not. Case three and case four represent that the communication events are the same and not, respectively. If the events are the same, then any one of these two snapshots can be added into the result trace; otherwise, both of these two snapshots should be added into the result trace.

Definition 9 (Parallel Composition on Networks).

$$\boldsymbol{beh}(M \| N) =_{df} \boldsymbol{beh}(M) \| \boldsymbol{beh}(N)$$

Finally, we give the semantics for the parallel composition of two networks. From Definition 9, we can see that the network behaviour of the parallel composition of networks can be transformed into the parallel composition of network behaviours of each network. Because the behaviour of a network is composed of network behaviours of each network nodes and we have already given the definitions of network behaviours of nodes in Subsect. 3.1, thus we omit the semantic details about the parallel composition of networks here. Note that, the denotational semantics for other processes can be found in Appendix B.

4   Algebraic Properties

Our work towards the formalization of mCWQ calculus for MANETs aims to deduce its interesting properties, which are usually expressed using algebraic laws and equations [9]. In this section, we explore a set of algebraic laws for mCWQ calculus, focusing on the set of parallel expansion laws for communications with binders. We use the notation $\rightarrow$ to represent that a possible first action may be executed in the algebraic laws.

(Par-1-1) Let $N_1 = n_1[c_i!v_i.P_1]_{T_1}^{f_1}$ $N_2 = n_2[\&_\forall(c_1?x_1, ..., c_i?x_i, ..., c_k?x_k).P_2]_{T_2}^{f_2}$

$N_3 = N_1 || N_2$, where $D(f_1(t), f_2(t)) \leq \text{Rad}(n_1, c_i)$.

Then $N_3 = c_i.v_i \rightarrow (n_1[P_1]_{T_1}^{f_1} || n_2[\&_\forall(c_1?x_1, ..., c_k?x_k).P_2\{v_i/x_i\}]_{T_2}^{f_2})$

In the law (Par-1-1), a node n_1 who is executing an output process, at some time t $(t \leq T_1)$, succeeds in broadcasting the message v_i over channel c_i and continues to execute the process P_1. Its location can be compute using its current mobility function f_1 as $f_1(t)$. All the other nodes executing an input process within the communication area of n_1, checked by the condition of $D(f_1(t), f_2(t)) \leq \text{Rad}(n_1, c_i)$, are able to receive that message. Suppose that one node n_2 is executing a receiving action with a binder whose quality predicate is $\forall$ at the location $f_2(t)$, which asks that all these k variables should receive messages before continuing to execute the process P_2. Thus, after receiving the message v_i via channel c_i, the quality predicate $\forall$ will be kept until the other $k-1$ messages are received. Note that, we use $c.v$ to stand for the communication event and $P_2\{v_i/x_i\}$ represents replacing x_i in the process P_2 by the message value v_i.

(Par-1-2) Let $N_1 = n_1[c_i!v_i.P_1]_{T_1}^{f_1}$ $N_2 = n_2[\&_\forall(c_1?x_1, ..., c_i?x_i, ..., c_k?x_k).P_2]_{T_2}^{f_2}$

$N_3 = N_1 || N_2$, where $D(f_1(t), f_2(t)) > \text{Rad}(n_1, c_i)$.

Then $N_3 = c_i!v_i \rightarrow (n_1[P_1]_{T_1}^{f_1} || n_2[\&_\forall(c_1?x_1, ..., c_i?x_i, ..., c_k?x_k).P_2]_{T_2}^{f_2})$

Comparing with (Par-1-1), the law in (Par-1-2) describes that a node n_1 who is executing an output process, at some time t $(t \leq T_1)$, succeeds in broadcasting the message v_i over channel c_i and continues to execute the process P_1, whereas all the other nodes executing an receiving action out of the communication area of n_1, checked by the condition of $D(f_1(t), f_2(t)) > \text{Rad}(n_1, c_i)$, cannot receive that message and will keep their original processes unchanged. Thus, the first action of the parallel composition is $c_i!v_i$.

Similarly, we will give the algebraic laws for the parallel composition between a sender and a receiver with binder, whose quality predicate is $\exists$ or m/k, respectively. Each parallel composition law should be divided into two cases, like

(Par-1-1) and (Par-1-2) based on whether the receiver is in the transmission area of the sender or not. However, due to the space limitation, we omit the case that the receiver is outside the transmission area of the sender in the following laws. The general idea for these cases is, like (Par-1-2), executing the broadcast action and remaining the processes inside the receiver unchanged.

(Par-2) Let $N_1 = n_1[c_i!v_i.P_1]_{T_1}^{f_1}$ $N_2 = n_2[\&_\exists(c_1?x_1, ..., c_i?x_i, ..., c_k?x_k).P_2]_{T_2}^{f_2}$

$\quad\quad N_3 = N_1 \| N_2$, where $D(f_1(t), f_2(t)) \leq \text{Rad}(n_1, c_i)$.

$\quad$ Then $N_3 = c_i.v_i \rightarrow (n_1[P_1]_{T_1}^{f_1} \| n_2[P_2\{v_i/x_i\}]_{T_2}^{f_2})$

The law (Par-2) describes the communication between the sender and the receiver with the quality predicate $\exists$ which requests that at least one variable should receive a message from the sender. After receiving the message v_i via channel c_i from the sender n_1, the quality predicate in the node n_2 is satisfied and it will continue to execute the following process P_2, using the message value v_i to replace the variable x_i. Similar with the law (Par-1-1), the sender n_1 will execute the following process P_1.

(Par-3) Let $N_1 = n_1[c_i!v_i.P_1]_{T_1}^{f_1}$

$$N_2 = n_2[\&_{m/k}(c_1?x_1, ..., c_i?x_i, ..., c_k?x_k).P_2]_{T_2}^{f_2}$$

$$N_3 = N_1 \| N_2, \text{ where } D(f_1(t), f_2(t)) \leq \text{Rad}(n_1, c_i). \text{ Then we have:}$$

$$N_3 = \begin{cases} c_i.v_i \rightarrow (n_1[P_1]_{T_1}^{f_1} \| n_2[\&_{(m-1)/k}(c_1?x_1, ..., c_k?x_k).P_2\{v_i/x_i\}]_{T_2}^{f_2}) \text{ if } m > 1, \\ c_i.v_i \rightarrow (n_1[P_1]_{T_1}^{f_1} \| n_2[P_2\{v_i/x_i\}]_{T_2}^{f_2}) \quad\quad\quad\quad\quad\quad\quad\quad \text{if } m = 1. \end{cases}$$

There is another quality predicate m/k that can be used in the receiving action with binder. It requires that at least m of k variables should receive messages so that the whole process can continue to execute the following process. In the law (Par-3), we take two cases into consideration: $m = 1$ and $m > 1$. Here, $m = 1$ means that only one variable is requested to receive messages, which has the same effect as the quality predicate equaling to $\exists$; otherwise, after receiving the message v_i via channel c_i, there are still $m - 1$ of k variables waiting for receiving messages from other senders.

5   Conclusion and Future Work

mCWQ calculus is motivated by the issue of wireless communication quality caused by node movements in MANETs. In this paper, we investigated the denotational semantics of mCWQ to provide a better understanding of MANETs, which also helped us to find more interesting properties of networks from mathematical perspective. Based on the UTP theory, the semantics of the whole network are expressed as basic commands, processes and parallel composition. We used an observation tuple and a trace variable tr to record the communications among wireless nodes as well as the time point when the communication happened. Besides, we also provided a set of algebraic laws related to the wireless communications with quality binders.

In the future, we will propose the guarded choice for mCWQ calculus and investigate more interesting properties based on it. Furthermore, exploring the linking theories between different semantics of mCWQ calculus is also interesting to be investigated.

Acknowledgements. This work was partly supported by National Natural Science Foundation of China (Grant No. 61872145), National Key Research and Development Program of China (Grant No. 2018YFB2101300), and Shanghai Collaborative Innovation Center of Trustworthy Software for Internet of Things (No. ZF1213).

A   Additional Definitions for Receiving with Binder

Definition 10 (Merging Rules for the Operator $\oplus$).

1. *If each of n traces is ε, then $tr_1 \oplus tr_2 \oplus \ldots \oplus tr_n =_{df} \{\varepsilon\}$*
2. *If $n-1$ of n traces equal to ε and one of n trace is not ε, denoted as $\langle(t, evt)\rangle\widehat{}s$, then $tr_1 \oplus tr_2 \oplus \ldots \oplus tr_n =_{df} \{\langle(t, evt)\rangle\widehat{}s\}$*
3. *If l of n traces equal to ε, whose indexes can be represented as $i_1, \ldots, i_l$, and k of n traces are not ε with indexes $j_1, \ldots, j_k$, denoted as $\langle(t_{j_1}, evt_{j_1})\rangle\widehat{}s_{j_1}, \ldots, \langle(t_{j_k}, evt_{j_k})\rangle\widehat{}s_{j_k}$, where $l \geq 0$ and $n = k+l$, then $tr_1 \oplus tr_2 \oplus \ldots \oplus tr_n =_{df}$*

$$\daleth_1\big(\widehat{first}\,(\langle(t_{j_1}, evt_{j_1})\rangle, \ldots, \langle(t_{j_k}, evt_{j_k})\rangle)\big)\widehat{}\big(\underset{cond}{\oplus}\langle(t_j, evt_j)\rangle\widehat{}s_j \oplus s_{j_i}\big)$$

where $i = \daleth_2\big(\widehat{first}\,(\langle(t_{j_1}, evt_{j_1})\rangle, \ldots, \langle(t_{j_k}, evt_{j_k})\rangle)\big)$ and
cond $= \forall j \in (\{j_1, \ldots, j_k\} - \{j_i\})$

From Definition 10, we can see that: (1) If all of these n traces are empty (represented by ε), the merge result of these traces is a singleton empty trace; (2) If $n-1$ of these n traces are empty trace and only one trace is non-empty trace, the merge result of these n traces is a set containing that non-empty trace; (3) If there are l ($l \geq 0$) traces that are empty traces and k ($k \geq 2$) traces that are non-empty traces, the merge result of these traces is given by a recursive function $\widehat{first}$ depending on the execution time of each trace. The definition of this recursive function can be found in Definition 11.

Definition 11 (Function for Searching the Earliest Executed Snapshot).

$$\widehat{first}\,(\langle(t_1, evt_1)\rangle, \ldots, \langle(t_n, evt_n)\rangle) =_{df}$$

$$\begin{cases} (\langle(t_1, evt_1)\rangle, 1) \\ \quad if\ \pi_1(\daleth_1(\langle(t_1, evt_1)\rangle, 1)) \leq \pi_1(\daleth_1(\widehat{first}\,(\langle(t_2, evt_2)\rangle, \ldots, \langle(t_n, evt_n)\rangle))), \\ \widehat{first}\,(\langle(t_2, evt_2)\rangle, \ldots, \langle(t_n, evt_n)\rangle) \\ \quad if\ \pi_1(\daleth_1(\langle(t_1, evt_1)\rangle, 1)) > \pi_1(\daleth_1(\widehat{first}\,(\langle(t_2, evt_2)\rangle, \ldots, \langle(t_n, evt_n)\rangle))). \end{cases}$$

We give the explanations for notations $\daleth_1$ and $\daleth_2$ used in Definition 10 and Definition 11 before describing the recursive function $\widehat{first}$. Both of these two notations are used on a data pair, which is composed of the trace snapshot and the index of this snapshot in the parameter list of the function $\widehat{first}$. Notation $\daleth_1$ denotes selecting the first component from the data pair, that is the trace snapshot; while notation $\daleth_2$ stands for selecting the second component from the data pair, that is the index of the trace in the parameter list of function $\widehat{first}$. Note that, notations $\daleth_1$ and $\daleth_2$ do a selection on the data pair, while notations π_1 and π_2 also do the selection, but on the snapshot of trace.

Recursive function $\widehat{first}$ takes n trace snapshots as its parameters and returns one data pair composed of the trace snapshot and its index, whose executing time is earliest among all trace snapshots. Firstly, the function $\widehat{first}$ uses $\daleth_1$ to get the first trace snapshot from its parameter list, then π_1 is used to get the executing time t_1 from this trace snapshot. Similarly, we apply the recursive function $\widehat{first}$ again on the remaining parameter list without the first trace snapshot, and use $\daleth_1$ and π_1 to obtain the executing time of the trace snapshot who is the earliest to be executed. If the earliest executing time of the remaining trace snapshots is not earlier than t_1, then the function $\widehat{first}$ returns the first trace snapshot of the parameter list and its corresponding index 1; otherwise, this function returns a pair of trace snapshot and its index, whose executing time is earliest, from the remaining $n-1$ trace snapshots.

B   Denotational Semantics for other Processes

In this section, we will describe the semantics for processes, including the inert process nil and *case* construction, as follows.

Since the execution of a nil process does not consume time, the interval of the observation time for its network behaviour equals to 0, and the execution state of the whole process is *terminate*. The execution trace, mobility function and its corresponding timeout are kept unchanged.

Definition 12 (Inert Process).

$$\boldsymbol{beh}(n[\text{nil}.P]_T^f) =_{df}$$

$$\mathcal{H} \left(\begin{array}{l} \overrightarrow{st} = terminate \ \wedge \ \overrightarrow{tr} = \overleftarrow{tr} \ \wedge \ \delta(time) = 0 \ \wedge \\ \overrightarrow{\mathcal{F}} = \overleftarrow{\mathcal{F}} \ \wedge \ \overrightarrow{\mathcal{T}} = \overleftarrow{\mathcal{T}} \ \wedge \ \overrightarrow{\mathcal{F}} = f \ \wedge \ \overleftarrow{\mathcal{T}} = T \end{array} \right) \ \fatsemi \ \boldsymbol{beh}(n[P]_{\mathcal{T}}^{\mathcal{F}})$$

In Definition 13, we give the denotational semantics for *case* construction. This operation is used to detect whether the receiving variable has already received values from channels, which is depending on the evaluation of the expression e. If the evaluation result equals to some data, then this data will be assigned to a data variable y and we can use y to replace the expression e in all free occurrences of e in process P. The network behaviour will behave like the behaviour of process P. Whereas, if the evaluation of the expression e equals to none, then the network behaviour will behave like the behaviour of process Q.

Definition 13 (*case* Construction).

$$beh(n[\text{case } e \text{ of } y : P \text{ else } Q]^f_T) =_{df} (\mathcal{B}(\|e\|) = \text{some}(v)) \wedge beh(n[P\{e/y\}]^f_T)) \bigvee$$

$$(\mathcal{B}(\|e\|) = \text{none}) \wedge beh(n[Q]^f_T))$$

Here, $\mathcal{B}$ is a boolean function, whose definition is given in Definition 14.

Definition 14. *Given a boolean expression Bexp, the boolean function $\mathcal{B}$ will return the truth value of the boolean expression, whose definition is : $\mathcal{B} : Bexp \rightarrow Boolean$.*

$$\mathcal{B}(true) \quad = true$$
$$\mathcal{B}(false) \quad = false$$
$$\mathcal{B}(exp_1 = exp_2) = \begin{cases} true & \|exp_1\| = \|exp_2\|, \\ false & \|exp_1\| \neq \|exp_2\|. \end{cases}$$

References

1. Butterfield, A., Sherif, A., Woodcock, J.: Slotted-circus. In: Davies, J., Gibbons, J. (eds.) IFM 2007. LNCS, vol. 4591, pp. 75–97. Springer, Heidelberg (2007). https://doi.org/10.1007/978-3-540-73210-5_5
2. Camp, T., Boleng, J., Davies, V.: A survey of mobility models for Ad Hoc network research. Wirel. Commun. Mob. Comput. **2**(5), 483–502 (2002)
3. Cavalcanti, A., Woodcock, J.: A tutorial introduction to CSP in *Unifying Theories of Programming*. In: Cavalcanti, A., Sampaio, A., Woodcock, J. (eds.) PSSE 2004. LNCS, vol. 3167, pp. 220–268. Springer, Heidelberg (2006). https://doi.org/10.1007/11889229_6
4. Djenouri, D., Khelladi, L., Badache, N.: A survey of security issues in Mobile Ad Hoc and sensor networks. IEEE Commun. Surv. Tutorials **7**(1–4), 2–28 (2005)
5. Fehnker, A., van Glabbeek, R., Höfner, P., McIver, A., Portmann, M., Tan, W.L.: A process algebra for wireless mesh networks. In: Seidl, H. (ed.) ESOP 2012. LNCS, vol. 7211, pp. 295–315. Springer, Heidelberg (2012). https://doi.org/10.1007/978-3-642-28869-2_15
6. Foster, S., Cavalcanti, A., Woodcock, J., Zeyda, F.: Unifying Theories of time with generalised reactive processes. Inf. Process. Lett. **135**, 47–52 (2018)
7. Godskesen, J.C.: A calculus for mobile Ad Hoc networks. In: Murphy, A.L., Vitek, J. (eds.) COORDINATION 2007. LNCS, vol. 4467, pp. 132–150. Springer, Heidelberg (2007). https://doi.org/10.1007/978-3-540-72794-1_8
8. Hoare, C.A.R., He, J.: Unifying Theories of Programming. Prentice Hall, Upper Saddle River (1998)
9. Hoare, T.: Laws of programming: the algebraic unification of theories of concurrency. In: Baldan, P., Gorla, D. (eds.) CONCUR 2014. LNCS, vol. 8704, pp. 1–6. Springer, Heidelberg (2014). https://doi.org/10.1007/978-3-662-44584-6_1

10. Lanese, I., Sangiorgi, D.: An operational semantics for a calculus for wireless systems. Theor. Comput. Sci. **411**(19), 1928–1948 (2010)
11. Lee, E.A.: Architectural support for cyber-physical systems. In: Proceedings og 12th International Conference on Architectural Support for Programming Languages and Operating Systems, ASPLOS 2015, pp. 14–18 (2015)
12. Merro, M.: An observational theory for mobile Ad Hoc networks (full version). Inf. Comput. **207**(2), 194–208 (2009)
13. Merro, M., Sibilio, E.: A timed calculus for wireless systems. In: Arbab, F., Sirjani, M. (eds.) FSEN 2009. LNCS, vol. 5961, pp. 228–243. Springer, Heidelberg (2010). https://doi.org/10.1007/978-3-642-11623-0_13
14. Mezzetti, N., Sangiorgi, D.: Towards a calculus for wireless systems. Electr. Notes Theor. Comput. Sci. **158**, 331–353 (2006)
15. Nielson, H.R., Nielson, F., Vigo, R.: A calculus for quality. In: Pǎsǎreanu, C.S., Salaün, G. (eds.) FACS 2012. LNCS, vol. 7684, pp. 188–204. Springer, Heidelberg (2013). https://doi.org/10.1007/978-3-642-35861-6_12
16. Oliveira, M., Cavalcanti, A., Woodcock, J.: A UTP semantics for circus. Formal Asp. Comput. **21**(1–2), 3–32 (2009)
17. Plotkin, G.D.: A structural approach to operational semantics. J. Log. Algebr. Program. **60–61**, 17–139 (2004)
18. Sherif, A., Jifeng, H., Cavalcanti, A., Sampaio, A.: A framework for specification and validation of real-time systems using *Circus* actions. In: Liu, Z., Araki, K. (eds.) ICTAC 2004. LNCS, vol. 3407, pp. 478–493. Springer, Heidelberg (2005). https://doi.org/10.1007/978-3-540-31862-0_34
19. Wei, K., Woodcock, J., Cavalcanti, A.: *Circus Time* with reactive designs. In: Wolff, B., Gaudel, M.-C., Feliachi, A. (eds.) UTP 2012. LNCS, vol. 7681, pp. 68–87. Springer, Heidelberg (2013). https://doi.org/10.1007/978-3-642-35705-3_3
20. Wu, X., Liu, S., Zhu, H., Zhao, Y.: Reasoning about group-based mobility in MANETs. In: Proceedings of 20th IEEE Pacific Rim International Symposium on Dependable Computing, PRDC 2014, pp. 244–253. IEEE Computer Society (2014)
21. Wu, X., Zhao, Y., Zhu, H.: Integrating a calculus with mobility and quality for wireless sensor networks. In: Proceedings of 17th IEEE International Symposium on High Assurance Systems Engineering, HASE 2016, pp. 220–227. IEEE Computer Society (2016)
22. Wu, X., Zhu, H.: A Calculus for wireless sensor networks from quality perspective. In: Proceedings of IEEE 16th International Symposium on High Assurance Systems Engineering, HASE 2015, pp. 223–231 (2015)
23. Zhu, H., Sanders, J.W., He, J., Qin, S.: Denotational semantics for a probabilistic timed shared-variable language. In: Wolff, B., Gaudel, M.-C., Feliachi, A. (eds.) UTP 2012. LNCS, vol. 7681, pp. 224–247. Springer, Heidelberg (2013). https://doi.org/10.1007/978-3-642-35705-3_11

Author Index

Printed in the United States
By Bookmasters